by Gary J. Lassiter

Evolution and Ungodly Ways

LCROSS Publishing
Boston

First edition: February, 2013

Publisher's Cataloging-in-Publication data   (from CIPblock.com)
Lassiter, Gary J.
    Evolution and ungodly ways / by Gary J. Lassiter.
    p. cm.
    Includes bibliographic references and index.
        ISBN 978-0-615-56364-0 *Hardback*
        ISBN 978-0-615-52282-1 *Paperback*
        Also available for *Kindle* and *Nook*.
1. Evolution (Biology). 2. Religion and science. 3. Evolution (Biology) --Religious aspects --Christianity. 4. Intelligent design (Teleology). 5. Natural history --Religious aspects. 6. Creationism. I. Title.
                    BS652 .L37 2011
231.7/652 --dc22   2011914433

LCROSS Publishing
Boston, Massachusetts
lcrosspublishing@gmail.com

To all my Christian friends, but most especially to my wife without whom this work would never have been possible and whose observations and contributions were essential to its completion. Any statements of appreciation from me could not be concluded without acknowledging the many helpful comments I have received from numerous readers, including David Bristol, Oded Haber, Gail Penrod, Adam Pierce, and Lee Teran.

Notes on the Text

Biblical quotations:

All biblical quotations in this book are taken from the New American Standard. While not the most popular English translation of the Bible, it is generally considered to be the most accurate.

Internet links:

The Reference Notes and Bibliography contain numerous links to useful material on the Internet. All of these links are accurate as of the date of the edition. Since the content of web pages can be changed, or moved to new locations, or temporarily taken off-line, or removed altogether, links can fail. Sometimes simply using a different browser can resolve the problem. In any case you may still be able to retrieve the original material by doing a search.

In electronic editions selecting these links will display the web pages. The success or failure of ebook links can depend on the device. A link which might be ignored or fail on an older or simpler device, most likely will succeed on a more capable ebook reader.

Contents

CONTENTS

Figures

Introduction

If you were looking for a reasonable discussion on the issues of science and religion, specifically evolution and Christianity, you have picked up the wrong book. Here you won't find any back and forth, no give and take, granting equal time and equal respect to differing points of view. Instead, this book is about betrayal and deceit; it's about the rewards gleamed from a successful distillation of enmity; it's about compromising one's character for an improved orchestration of lies; it's about nurturing contempt for one's opponents with mixed measures of fear, arrogance and desperation; it's about the vast powers of denial and a corruption of intellectual thought so convicted that an elimination of knowledge is repeatedly greeted with applause; but most of all this book is about a phenomenon of self-destruction brought to the public stage by a widespread deterioration, if not outright abandonment, of all the benefits of even the most casual discernment. If this is not the kind of book you were looking for, yet you've gotten this far, maybe you should try a few more pages.

Evolution
and Ungodly Ways

1

A Spirit of Error

From the first line in the Bible, God created the heavens and the earth. Scientists investigate the nature of this creation. In the light of its biblical explanation, whatever they discover must be a part of God's design. The theory of relativity, for example, is a discovery about some of the physical characteristics of the universe and must be, therefore, one of the components of the divine creation. If evolution is, as scientists assert, a fact of nature, then it too must be a part of God's vast and intricate plan. Assuming this to be the case, then Christians who deny evolution are taking a stand against the work of God. Believers in Christ who walk a path in opposition to God behave in ungodly ways. Is there any evidence that Christian evolution-deniers behave in ungodly ways?

- Do they set aside familiar virtues of Christianity, such as honesty, repentance, humility, and respect and express instead the counter characteristics of dishonesty, un-repentance, arrogance, and scorn?

- Do they withhold the whole truth, defend instead a contrivance of half-truths, using false teachings to conduct educational and propaganda campaigns of misinformation and misinterpretation?

- Do they create discord with non-believers and seek power and influence through the cultivation of ignorance and animosity among their fellow Christians?

If abandoning Christian values and deceiving the faithful are common attributes and goals of Christian evolution-deniers, then the employment of this behavior exposes evolution-denial, both personally and communally, as one of the great self-destructive forces in Christianity today.

2

Nothing as Advertised

In books and in videos, from monologues and dialogues on the Internet to speeches and live debates and even testimonies under oath, Christian evolution-deniers disseminate misinformation about science as though it were a calling. Here is Emeritus Professor of Law Philip E. Johnson, the "godfather" of modern intelligent design and author of *Darwin on Trial*, teaching his fellow Christians a fantasy about what the fossil record is supposed to look like:

> Darwin's theory predicted not merely that fossil transitionals would be found; it implied that a truly complete fossil record would be mostly transitionals, and that what we think of as fixed species would be revealed as mere arbitrary viewpoints in a process of continual change.[1]

In fact, "Darwin's theory" neither predicts nor implies anything even remotely like Johnson's description. Here is Charles Darwin himself:

> All these causes taken conjointly, must have tended to make the geological record extremely imperfect, and will to a large extent explain why we do not find interminable varieties, connecting together all the extinct and existing forms of life by the finest graduated steps.[2]

Professor Johnson's false description is made all the worse by the attempt at cleverness. Switching verbs from "predicted" to "implied" in a crafty sentence structure, he presents an imaginary "truly complete fossil record" as though it were an expected part of reality. Carrying this further and drawing conclusions from this false reality about what "would be revealed" is deceit.

In a brief but famous review of *Darwin on Trial*, paleontologist Stephen Jay Gould pointed out a river of errors flowing through the book.[3] There are errors of fact, information out-of-date by decades, hackneyed canards, half-truths, and in abundance the lawyer's speciality of rhetorical trickery, as in the above example.

In Attorney Johnson's second edition of his book, published two years after the first, he took note of Gould's critical review, as well as those of other scientists, and corrected exactly *one* error.[4] All the other errors, including the one above, were retained. Apparently they are all actually meant to be there.

Some of Johnson's critics have complained about the

length a thorough review would take to explain the many errors in *Darwin on Trial* and in other books Johnson has written.[5] But Professor Johnson easily dealt with all such assessments when he simply swept them off the table, stating:

> Gould listed a string of objections about matters that had nothing to do with the main line of argument.[6]

Evasive and dismissive retorts such as this only serve to accentuate the deterioration of Christian influence. Real leadership is a vision standing on a foundation of truth. But when the foundation is built out of layer upon layer of lies upon lies, the leader becomes just another sycophant serving the deception. For those who expect Christians to be as advertised, honest and without guile, they will have to find some other literature to explore than the many Christian works on evolution.

3

Self-Destruction

Part I: *A Classic Story of Natural Selection*

In the early decades of the nineteenth century, shortly after the start of the industrial revolution, soot pollution from the new coal-burning economy coated urban areas with a grimy stain – a discoloration which also spread across the surrounding countryside, darkening the landscape over large regions. Even the sky was painted with a persistent smokey haze. But a little more than a hundred years later, in the middle of the twentieth century, changes to the industrial base, including the dispersion of industry, a shift to different energy sources, and the implementation of cleaner air standards, eliminated the source of pollution. Subsequent biological growth along with a natural washing away of the environmental blemish restored the tainted regions to their original lighter tones. This geographic color cycle occurred in many locations around the industrialized world – first in the British Isles, but soon thereafter in other areas of Europe, North America, and Japan.

Coincident with the initial burst of industrialization in Great Britain was a rising interest in the science of biology, including of course entomology – the study of insects. As various regions went through this man-made environmental gyration, it was observed that some local insect populations, most prominently peppered moths and two-spot ladybugs, changed from lighter colored majorities to darker ones, and then back again.[7] In particular, the first dark colored peppered moth was reportedly captured in 1848. But its novelty was soon lost as light colored moths vanished to such an extent that by 1895 the population of moths in and around Manchester, England was 98% dark colored.[8] However, with the restoration of the original lighter tones to the landscape, the peppered moths have been reverting to their original lighter colored camouflage. These transformations to the moth population have been attributed to the fact that birds prey more often on more easily seen moths.

The story of the peppered moth is taught in almost every introduction to biology as a rare example of evolution occurring within a single human lifetime. That an environmental change (industrial soot pollution) can lead to alterations in the characteristics of a species (peppered moth camouflage) fits perfectly within the explanatory power of the theory of evolution wherein a process of natural selection (bird predation) provides the driving force to transform the population. The fact that these moths are currently returning to their original state, after the reversal of the environmental change, nicely accentuates the credibility of the theory. An enormous accumulation of diverse and consistent evidence, published in thousands of pages of scientific journals, supports the evolutionary explanation while there is no evidence in support of any other explanation.[9] As the late peppered moth expert Dr. M. E. N. Majerus said:

> If the rise and fall of the peppered moth is one of the most visually impacting and easily understood examples of Darwinian evolution in action, it should be taught. It provides after all the proof of evolution.

> The peppered moth story is easy to understand, because it involves things that we are familiar with: vision and predation and birds and moths and pollution and camouflage and lunch and death. That is why the anti-evolution lobby attacks the peppered moth story. They are frightened that too many people will be able to understand.[10]

Part II: *Destroying Darwinism*

Having gotten his first Ph.D. in Religious Studies, Dr. Jonathan Wells decided to get a second Ph.D. in Molecular and Cell Biology specifically, as he has famously stated, to "devote my life to destroying Darwinism."[11] Here, from his book *Icons of Evolution: Science or Myth? Why Much of What We Teach About Evolution Is Wrong*, is Dr. Wells writing on the peppered moths:

> Most introductory textbooks now illustrate this classical story of natural selection with photographs of the two varieties of peppered

moth resting on light- and dark-colored tree trunks . . . What the textbooks don't explain, however, is that biologists have known since the 1980's that the classical story has some serious flaws. The most serious is that peppered moths in the wild don't even rest on tree trunks. The textbook photographs, it turns out, have been staged.[12]

A few pages later Dr. Wells continues down this path. Since, according to his assurances, peppered moths do not reside on tree trunks:

> Pictures of peppered moths on tree trunks must be staged. Some are made using dead specimens that are glued or pinned to the trunk, while others use live specimens that are manually placed in desired positions. Since peppered moths are quite torpid in daylight, they remain where they are put.
>
> Manually positioned moths have also been used to make television nature documentaries. University of Massachusetts biologist Theodore Sargent told a *Washington Times* reporter in 1999 that he once glued some dead specimens on a tree trunk for a TV documentary about peppered moths.
>
> Staged photos may have been reasonable when biologists thought they were simulating the normal resting-places of peppered moths.

By the late 1980s, however, the practice should have stopped. Yet according to Sargent, a lot of faked photographs have been made since then.[13]

Among the authoritative books on this subject, and one which Dr. Wells references in his book, is *Melanism: Evolution in Action* by the aforementioned M. E. N. Majerus. Having studied these moths for decades, Professor Majerus provided the actual scientific data on their resting places in trees. Over a period of 32 years, observations of 47 moths in the wild found:[14]

15 in the branches
20 on the trunk/branch boundary
12 on the trunks

Given the references Dr. Wells makes to Majerus' book, he had this information in hand when he wrote his own book.

Although Majerus considered the above data "pitifully scarce," with less than 50 wild moths observed in their natural setting over more than three decades, he nevertheless made it clear that bird predation of the light and dark moths, "in habitats affected by industrial pollution to different degrees, is the primary influence on the evolution of melanism in the peppered moth."[15] It might be worthwhile to note that since the publication of Dr. Wells' book in 2000, Majerus, from observations made over a six year period beginning in 2001, verified and expanded on the data set by observing a total of 135 wild peppered moths, of which 50 were resting on tree trunks.[16]

While Dr. Wells' tree trunk complaint is just plain false,

his second complaint, that the textbook photographs are "staged," is just plain deceitful. Why wouldn't they be staged? It's a textbook. Educational books, and "television nature documentaries," often contain staged, illustrative photographs.

According to Majerus, none of the photographs of peppered moths in his book were staged and, as anyone who cares to do a comparison can see, they look exactly like the staged photos. But even if the textbook photographs were copied straight from scientific papers published in peer-reviewed scientific journals wherein the scientists themselves had actually glued the moths to the tree trunks, it doesn't matter because photographs of these moths have *never* constituted the scientific data from which any conclusions have been drawn. You can find an example of real scientific data on the opposite page in the table provided by Majerus.

Over the years those scientists who did in fact conduct experiments by glueing moths to tree trunks, for example to study how the density of moths affects the predatory behavior of birds, did nothing to misrepresent the central and dramatic fact – namely that natural selection, in this case driven by the hunting skills of hungry birds, repeatedly transformed a population of moths, first to dark colored and then back to light colored, as their environment changed with the rise and fall of industrial soot pollution, in exact agreement with the theory of evolution. Dr. Wells' fixation on scientists and educators glueing moths to tree trunks merely serves his purpose of guiding his readers away from this essential truth. By harping on an irrelevance while ignoring thousands of pages of the complete scientific literature on this little corner of nature, including that page with Majerus' table of resting positions, Dr. Wells hopes to accomplish his mission of destruction, plainly by any means language and manipulation can provide.

To repudiate Dr. Wells' disingenuous and incomplete presentation of the peppered moth data, and to clarify the significance of these moths, here again is Majerus speaking for all the data and all of his colleagues:

> The critics of work on this case and those who cast doubt on its validity are, without exception, persons who have, as far as I know, never bred the moth and never conducted an experiment on it. In most cases they have probably never seen a live Peppered moth in the wild. Perhaps those who have the most intimate knowledge of this moth are the scientists who have bred it, watched it and studied it, in both the laboratory and the wild. These include, among others, the late Sir Cyril Clarke, Professors Paul Brakefield, Laurence Cook, Bruce Grant, K. Mikkola, Drs Rory Howlett, Carys Jones, David Lees, John Muggleton and myself. I believe that, without exception, it is our view that the case of melanism in the Peppered moth still stands as one of the best examples of evolution, by natural selection, in action.[17]

It is difficult picking the single most deceitful statement made by Dr. Wells on the recent evolution of the peppered moth, but the following certainly deserves to be in the top ten:

14

A critic of Darwin's theory might object that this "most spectacular evolutionary change ever witnessed" falls far short of providing a sufficient mechanism for evolution. After all, the only thing that happened was a change in the proportion of two varieties of a pre-existing species of moth. Although the change was dramatic, it is no more impressive than the changes domestic breeders have produced for centuries.[18]

Unfortunately, Dr. Wells fails to mention the most significant fact of all – that when the environment changed, it was nature that changed the population of moths. There is no evidence that God changed the moth population, nor any evidence that an unspecified, invisible, super powerful intelligent designer conducted the changes, but there is a library full of scientific publications cataloguing the evidence that nature changed the moth, accompanied by repeating evidence that Christian evolution-deniers maintain a dishonesty about the subject.[19]

But more than one nefarious dimension is revealed in the above quote from Dr. Wells. Darwin begins his scientific masterpiece by introducing his readers to his earliest thoughts on the subject. In his usual lengthy fashion, he enumerates the many successes domestic breeders have had with a broad range of plants and animals. From this Darwin raises the idea that if mankind, by careful and deliberate selections, can so easily and in so short a time, take such great advantage of a variability built into "organic beings," from cabbages to cows, then imagine how much more nature can accomplish, even down to the smallest refinement, given the enormous amount of time available across the history of life as well as the huge amount of stock with which to work. Darwin summed it up this way:

> Why, if man can by patience select variations most useful to himself, should nature fail in selecting variations useful, under changing conditions of life, to her living products? What limit can be put to this power, acting during long ages and rigidly scrutinising the whole constitution, structure, and habits of each creature, –favouring the good and rejecting the bad? I can see no limit to this power, in slowly and beautifully adapting each form to the most complex relations of life.[20]

In the years since the publication of *Icons of Evolution*, Dr. Wells has not corrected his errors on the peppered moths, nor has he corrected any of the many other errors which can be found in every chapter of his book.[21] The chapters, the "Icons," line up like a row of funhouse mirrors at a carnival, each presenting its own elaborate aberration of the truth. As with Philip Johnson before him, the errors, it seems, are all actually meant to be there. Clearly their hope is to convince themselves and their public that they have provided an image of reality which is honest, all the while expressing confidence that their supporters will never complain about how the distortions reflect on them.

To the astonishment of anyone even modestly acquainted with science, the falsehoods, manipulations, inaccuracies and denials radiating from the luminaries of Christian anti-evolution are so blatant and so widespread, only a concerted commitment to deceit can sustain them. For evolution-deniers the whole truth constantly gets in the way and is detested to the point where lying about science has become a mission of real passion. In contempt of their own professed values, they lead their followers into the shadows of an embarrassing ignorance, all the while promoting this journey as a sensible path. Their achievements, however, repeatedly testify as to how easily a group of people can promote the destruction of the very ideals they claim to care about the most.

4

Suspending More Than Disbelief

Propagandists always seem to succumb to a common delusion – a self-serving diminished estimate of the intelligence of ordinary people. Add to this the feedback of support they get when playing to a crowd of like-minded followers and pretty soon the crude sleight of hand is admired for its finesse, a clumsy distortion is offered up as though it had some real elegance, and outlandish assertions are made as though no one will notice they are actually nailed to a vacuum.

In an article titled, "Teach the Controversy on Origins," Stephen C. Meyer, with a Ph.D. in the Philosophy of Science from Cambridge University, senior fellow and vice president of The Discovery Institute, and the program director of its Center for Science and Culture, commented on his appearance in early March of 2002 before the Ohio State Board of Education where he expressed his concern for the proper education of high school students:

> While testifying before the state board, biolo-
> gist Dr. Jonathan Wells and I, submitted an
> annotated bibliography of over 40 peer-
> reviewed scientific articles that raise signifi-
> cant challenges to key tenets of Darwinian
> evolution. If students are to be required to
> master the case for Darwinian evolution (as
> we think they should), shouldn't they also
> know some of the difficulties described in
> such scientific literature?
>
> Shouldn't students know that many scientists
> doubt that the overall pattern of fossil evi-
> dence conforms to the Darwinian picture of
> the history of life? Shouldn't they know that
> some scientists now question previously
> stock Darwinian arguments from embryology
> and homology? And shouldn't they also
> know that many scientists now question the
> ability of natural selection to create funda-
> mentally new structures, organisms and body
> plans?[22]

With the theory of evolution holding a tenacious grip on
the academic community, and with its extensive influence
found throughout the vernacular of modern society, the
intelligent design rallying cry of "Teach the Controversy"
needed some solid measure of scientific support, some rock
of justification upon which to stand, and sure enough, there
it was – they said – in more than 40 esoteric, state-of-the-art

scientific publications. In their now infamous *Bibliography of Supplementary Resources for Ohio Science Education,* Dr. Meyer and Dr. Wells provided the following preface:

> The publications represent dissenting viewpoints that challenge one or another aspect of neo-Darwinism (the prevailing theory of evolution taught in biology textbooks), discuss problems that evolutionary theory faces, or suggest important new lines of evidence that biology must consider when explaining origins.[23]

The National Center for Science Education (NCSE), an organization of several thousand teachers, scientists, clergy and concerned citizens dedicated to defending the teaching of evolution in public schools, decided to investigate Dr. Meyer's bibliography of "significant challenges" and "dissenting viewpoints" and the doubts of "many scientists" by contacting directly the authors of the forty four articles.[24] Twenty six of the authors, representing thirty four of the publications, responded to the NCSE's questionnaire. Contrary to Dr. Meyer's pronouncements, not a single scientist considered their research to exhibit any evidence against the theory of evolution. Here are some of the comments by the scientists themselves:

> I state clearly that evolution is beyond dispute based on all the evidence I am aware of.
> *Kenneth Weiss*

they have selectively quoted just parts of what I wrote in order to distort completely what I said in the article.

Rodney Brooks

In no way does my work represent an attack on the theory of descent with modification, i.e. the plain fact of evolution, or even the fundamental insights of the neo-Darwinian theory of evolution.

Günther P. Wagner

[My paper is evidence for] the effectiveness of evolution in fine-tuning the properties and features of natural systems.

Philip Ball

I don't think it is a good representation of our work – our work does not present "a classic challenge to evolutionary analysis."

Peter J. Lockhart

My research on turtles and my research into evolutionary developmental biology is fully within Darwinian parameters.

Scott F. Gilbert

The words enclosed in quotation marks are accurate. However, the quotes are entirely misinterpreted and taken out of context.

David P. Mindell

While the article considers the relationship between micro- and macro- evolution, the Discovery Institute is inaccurate in saying that I am challenging the standard view of evolution. The treatment of macroevolution in that paper is an extension, but by no means a challenge.

Douglas L. Erwin

. . . the conclusion that this is "a hypothesis quite unexpected on neo-Darwinian (common ancestry) assumptions" is (i) not taken from our paper and (ii) not at all compatible with the data or ideas presented in the paper.

Eugene V. Koonin

Partly accurate and partly ambiguous. The creationists have taken a very complicated argument and extracted from it the bits and pieces that fit their world view.

Michael K. Richardson

The implications, particularly that molecular

data are unable to reconstruct the history of
life, are complete distortions of what we said.
Paul Morris

How can the ambition to hijack these papers be explained
unless the tenets of evolution-denial supersede even the most
basic principles of Christian doctrine? Evolution-deniers
present a unique theater. After taking your seat, you're
required to not only suspend your disbelief, but your values
as well.

5

Placing Your Trust in the Wrong Mathematician

When we define something we get to choose between two different approaches – we can either provide a positive definition or a negative definition. Here is an example of a positive definition, in this case, of a Christian:

> A Christian is someone who accepts Jesus as their Lord and Savior.

On the other hand, here is an example of a negative definition of a Christian:

> A Christian is not a Hindu, and not a Buddhist.

Both of these definitions are true, yet there is a huge difference between them.

The positive definition is a list of elements which are all related to each other. From this one simple sentence we learn that a Christian is a person who believes in Jesus for their salvation, for their life everlasting, while expressing a willingness to obey His leadership. It's basically the Good News in a single sentence.

In the negative definition, however, none of that appears. There is no mention of Jesus, obedience, or salvation. If this negative definition were applied worldwide, all Muslims would pass through the filter and be defined as Christians. To correct this error, we should add "and not a Muslim" to the list. But that still leaves Jews, Taoists, atheists, Sikhs, the practitioners of Shinto, Rastifarians, Jains, neo-Pagans, Zoroastrians, agnostics, and so on – all defined as Christians. But worse than that, if you look closely at the negative statement you will see that there isn't a single positive component. In particular there is no reference to a Christian being a person, thereby allowing cabbages, cows, the color orange, humming, igneous rocks, demons, and the feeling of astonishment to also pass through the filter and be defined as Christians. Clearly the list of negative elements is incomplete, which makes the statement invariably unreliable in accurately defining and identifying a Christian. But even if the list could somehow be completed, in the end, a negative definition would never say anything about what a Christian actually is.

Although it's obvious both of these definitions are true statements, if you had to choose between them, which would you pick as the better one – the positive definition or the negative?

Now suppose we make a list of all the equations of science from ancient geometry to modern relativity, like this:

> The circumference of a circle is *pi* times its diameter.
>
> .
>
> .
>
> .
>
> Energy is mass times the speed of light squared.

Feel free to fill in the rest, but notice that every equation is a *positive* definition and, as such, each reveals universal relationships among the fundamental constituents of our reality. This compact accumulation of scientific knowledge, gathered across thousands of years by gifts of great insight, is a treasure chest of immeasurable value. If we were to replace any of these equations with a *negative* definition, the loss of knowledge would be spectacular, for example:

> The circumference of a circle is not an area, and it is not an angle.

William A. Dembski, with a B.A. in Psychology, a M.S. in Statistics, a Ph.D. in Mathematics, another Ph.D. in Philosophy and a Master of Divinity degree from Princeton Theological Seminary, a senior fellow of The Discovery Institute's Center for Science and Culture, author of numerous works on intelligent design, and famous for having been extolled by his fellow evolution-deniers as "the Isaac Newton of information theory," claims to be able to detect the existence of intelligent design in our world with his "Explanatory Filter."

Here it is:

> Design is not regularity (i.e. laws of nature),
> and it is not chance.[25]

William Dembski wishes to be the first person in history to put a negative definition into the treasure chest of scientific knowledge. Scientists and mathematicians, however, are not so easily scammed.

Usually Dr. Dembski's definition of design is repackaged like this:

> Design is contingency and complexity and specification.

A switch to impressive new terminology with no apparent negative components now appears to present a positive definition! But rhetorical trickery is always a thin disguise. Pull aside the veil and you will discover that "contingency" is defined as an event or object not attributable to the laws of nature, and both "complexity" and "specification" are defined as events or objects not attributable to chance.[26]

Not only is this definition of design profoundly flawed because it's a negative definition – it is also a *failure* of a negative definition. According to Dr. Dembski, design is not a product of the laws of nature. But we don't know what all of the laws of nature are. We only know some of them. The whole purpose of science is to figure out what they are and that work is very far from complete. Design is therefore something which is not derived from something that we don't fully understand. Even our negative definition of a Christian is superior to this.

It is not difficult finding examples of just how deeply flawed Dembski's Explanatory Filter is. The most obvious reside in our past when previous generations, with less knowledge of the laws of nature, looked at the world through this filter and saw an intelligent agency behind almost every unusual event or object, including comets, droughts, albinos, storms, floods, meteors, fires, epidemics, lightening strikes, earthquakes, variable stars, epilepsy, lodestones, St. Elmo's fire, novas and supernovas, chemical reactions and eclipses. Least you think false assignments of natural events to some intervening intelligence are a thing of the past, there are contemporary examples.

In 1967 two radio astronomers discovered the first pulsar (pulsating star) and initially, though admittedly not very seriously, they labeled it LGM-1 for "Little Green Men."[27] Considering everything else anyone had ever observed with a radio telescope, the pulsations they detected were so improbably regular that they looked exactly like a signal from an intelligently designed transmitter. Within a year, however, pulsars were theorized to be rotating neutron stars which formed during the gravitational collapse of the core of exploding stars. This new "law of nature" was confirmed when a pulsar was discovered exactly where it was predicted to be found – in the center of the Crab Nebula, which is the debris of a star known to have exploded in the year 1054. The improbability of the signal's regularity, coupled with the fact that initially there was no natural explanation for such a radio signal from interstellar space, led to a somewhat facetious presumption of intelligent design where in fact there was no such design.

A much more dramatic and disturbing failure of Dembski's Explanatory Filter occurred in recent years due to an appalling miscarriage of justice in one of the world's most advanced countries. Several women in Great Britain were jailed for life for the incomprehensible crime of repeatedly murdering their newborns.[28] In each case the first such death was easily acknowledged as a devastating tragedy most probably caused by the poorly understood and very rare Sudden Infant Death Syndrome (SIDS). But suspicions arose when a tragedy so rare happened twice in the same family. By the third occurrence the prosecution's success could not have been easier. With no known natural cause of SIDS, plus the astronomical improbability of it happening repeatedly to the same people, it becomes effortless to deduce murder, which is death by design. But then science pushed back the shadows of ignorance just a bit by discovering the possibility that these deaths may have been due to a law of nature, specifically genetic factors which might make it much more likely for such multiple tragedies to happen in the same family. This scientific revelation helped to free these tormented women. Not knowing all of the laws of nature once again led to a determination of design where there was no design.

Imagine how much worse our situation would be if, instead of acquiring more knowledge about the laws of nature, we actually sought to eliminate some of that knowledge – say, for example, knowledge of the scientific theories and their supporting evidence explaining the natural origins of pulsars, or the causes of SIDS, or the origins of biological species. We would then be even more likely to falsely detect design.

Dr. Dembski's Explanatory Filter is, to put it simply, a construction of rhetorical trickery, fundamentally flawed, untrustworthy, and therefore useless. While his professional

efforts have been met with derision from his fellow mathematicians, Dr. Dembski has nevertheless achieved the
distinction of being perhaps the only mathematician in
history to have inspired, for quite a long time now, a set of
seemingly real seminars in statistical mathematics which are
presented only at religious gatherings by independent
pseudo-mathematician volunteers who possess no real
understanding of what they think they are communicating.
If Dr. Dembski has proved anything during his career as a
mathematician, it is that Christians can be easily scammed.
In fact, you could almost believe they want to be.

6

Not Good Enough

When biochemist Michael J. Behe studied the molecules making up the components of living cells, he found it hard to believe that nature could produce anything so extraordinarily intricate. So he didn't believe it. In 1996 he published *Darwin's Black Box: The Biochemical Challenge to Evolution* and introduced the concept of irreducible complexity.[29]

Dr. Behe noticed that all sorts of mechanisms, from the simple mousetrap to the not-so-simple human immune system, are made up of a number of unique and essential parts. Any mechanism, reduced to only its essential parts, is not functional if reduced any further. A half-built mousetrap will catch no mice.

According to Professor Behe and his evolution-denying compatriots, living organisms cannot possibly evolve new capabilities because evolution, in building up its creations one small, random improvement at a time, has insurmountable problems with the complexity of so many unique and essential components, all of which must be in place before any mechanism can perform its function. First, there is the sheer mathematical improbability of a series of essential parts being created one after the other in the proper order by random acts of nature. Second, there is the requirement that existing parts not become damaged by mutations while waiting for the next essential part to appear. And third, any nonfunctional, half-built molecular mechanism would grant no advantage to the organism, but would instead be a drain on resources and, as such, would most likely be weeded out by evolution long before the next random mutation could create the next unique part. Dr. Behe surmised that the theory of evolution contains a fatal flaw which had been missed by all other scientists for one hundred and fifty years. It seems the only solution to the existence of complex molecular systems in living cells is the creation of all those unique and essential parts by a supremely powerful intelligent designer.

There are many examples of the falsehoods and flaws in Professor Behe's observations and conclusions regarding his concept of irreducible complexity. Here are just a few of them:

- Though conceiving his initial ideas in 1912, it wasn't until 1918 that biologist Herman Muller published a paper explaining that evolution would naturally create systems which fit the description of what Dr. Behe calls "irreducible complexity." Muller referred to "the interlocking action of very numerous different elementary parts" and described how complicated molecular mechanisms with unique and essential components would inevitably occur from ordinary random mutations – see the simple three-step process described on the next page.[30]

- There are notable allegorical failures of irreducible complexity, as in the pretend, step-by-step "evolution" of a mousetrap to counter Dr. Behe's symbolic mousetrap analogy – see the referenced web sites by Alex Fidelibus and by Professor John H. McDonald.[31]

- There are examples from the fossil record of complex, interlocking structures being slowly created over time by random mutations, such as the gradual evolution of the three tiny, intricately specialized bones found in the human middle ear beginning around three hundred million years ago in the back of the lower jaw of ancient reptile-like animals.[32]

- There are molecular examples of the gradual, step-by-step creation of extremely complex mechanisms, including systems which can only be described as "overly complex," such as the inefficient, badly designed, unnecessarily long cascade of events which take place during blood clotting.[33] In fact, if you were looking for a purely natural, completely mindless "complexity generator" which could create perfectly workable, but occasionally overly complex mechanisms, then the simple biological-molecular process described below accomplishes all of that.

In evolution, where one small improvement at a time takes place, it's easy to create what appears to be an irreducibly complex molecular system. In fact, as Muller pointed out long ago, it's unavoidable. Here is an example, in only three steps, in only three mutations, using the oxygen transport molecule hemoglobin.

Step 1:
Floating around in the cell is a molecule we'll call X. It has nothing to do with hemoglobin. During cell division a mistake occurs and the DNA which codes for X gets copied twice. This is a well known phenomenon called gene duplication. Now we have two genes making molecule X.

Step 2:
Since the duplicated gene does no harm, it gets distributed in the population through the passage of many generations. At any moment in time some individuals have it and some don't. Eventually a mutation occurs in one of these duplicated genes, causing it to produce a new, unique molecule Y. Molecule X is still being produced by the other gene. As it turns out molecule Y just happens to stick to hemoglobin and makes hemoglobin work a little better. The new molecule is not essential. It just helps out. A person can live fine without Y, but those who have it carry a slightly better performing hemoglobin in their blood. Since this mutation confers a slight advantage, it survives well in the population.

Step 3:

Many generations later, with the nonessential but helpful Y mutation now scattered throughout the population, another mutation occurs, this time to the DNA which makes the hemoglobin molecule. This new mutation causes hemoglobin to change shape, part of it folding over onto Y, resulting in a combination which works much, much better than anything before. However, without Y in place the electrical forces around the molecule are such that the fold will go the wrong way, producing a hemoglobin disaster. At this point Y has become essential. The new mutation in the hemoglobin DNA *must* be accompanied by the gene which produces the unique molecule Y. After many generations the majority of the population will have the new combined hemoglobin-Y molecule in their blood because it is the top performer.

The outcome of these three steps is that (i) the hemoglobin molecule now does its job a lot better, and (ii) the system fits the description of irreducible complexity. But it got there via an evolutionary process based on a simple set of random mutations and not by the craftsmanship of any intelligent being.

The same process which improves the hemoglobin molecule by adding the product of a single modified gene can also be used to improve a large, elaborate assembly of molecules by adding a fortuitously modified part which is itself hugely complex. The evolution of amazingly complex molecular mechanisms is not so far fetched at all. Since new improvements build upon a foundation of previous improvements, the earlier modifications can easily become essential and unique parts. As Herman Muller explained long before Michael Behe was even born, the appearance of irreducible complexity is just another example of evolution at work.

How easy is it for Professor Behe to acknowledge that an evolutionary explanation can be a valid description of reality? In his decision declaring it unconstitutional to teach intelligent design as a scientific subject in a publicly funded school, Federal Judge John E. Jones III wrote:

> However, Dr. (Kenneth) Miller presented peer-reviewed studies refuting Professor Behe's claim that the immune system was irreducibly complex. Between 1996 and 2002, various studies confirmed each element of the evolutionary hypothesis explaining the origin of the immune system. In fact, on cross-examination, Professor Behe was questioned concerning his 1996 claim that science would never find an evolutionary explanation for the immune system. He was presented with fifty eight peer-reviewed publications, nine books, and several immunology textbook chapters about the evolution of the immune system; however, he simply insisted that this was still not sufficient evidence of evolution, and that it was not "good enough."[34]

It is hard to imagine a higher level of arrogance than a biochemist who relentlessly insists that a completely unpredictable, incomprehensible, and unmeasurable "intelligent designer" should be acknowledged by science in spite of massive evidence indicating that such an assertion is scientifically both groundless and unnecessary. Scientists work in a very free-thinking environment where skepticism is professionally encouraged. They are, therefore, not unfamiliar with occasional committed contrarians. They simply remain attentive to the evidence and are thereby not fooled by individuals failing to seek the truth from all the facts.

Christians, on the other hand, come from a very different point of view where God's supernatural revelations illuminate the truth. Moses could not fail to recognize the truth of God in the supernatural impossibility of the burning bush. The Jews escaping from Egypt could not fail to recognize God in the impossibility of that path across the bottom of the Red Sea. The very first Christians came face to face with the impossibility of Christ alive again. Christians easily identify with the Apostle Paul's supernaturally interrupted journey on the road to Damascus. The experiences of trusted fellow Christians and one's own personal knowledge of answered prayers as well as God's even most minor interventions all testify for Christians to the living presence of their Lord.

In all of this however, revelations of God are always limited to specific people at specific times. There is no example of any of God's supernatural acts remaining an ever-present reality in our natural world. If such a manifestation were to reside among us, there would be no need for faith, and certainly no reason for the Bible to repeatedly mention the importance of faith. "Walking in faith" becomes an unnecessary endeavor when the reality of God is available for all to see at the other end of the microscope, just turn the knob to bring it into focus.

So here also it is hard to imagine a higher level of arrogance than a biochemist who relentlessly insists he has made a discovery which reveals impossible things in spite of extensive and profound biblical evidence that God has never created anything in this "fallen world" which is both impossible and ever-present. These two characteristics can only be found in places where people are *not* separated from God, namely at the beginning in Eden, and at the end in the New Jerusalem.

The promoters of evolution-denial help their admirers to forget that throughout a consistent history spanning many thousands of years, just as there is no other-worldly burning bush available on some mountaintop for all to see, so also there is no impossible molecule announcing the presence of the Lord to anyone and everyone.

7

Torched by Thermodynamics

A strawman is a fake man. By creating strawman renditions of scientific explanations, evolution-deniers mislead their Christian audiences, captivating them with the impossibility of evolution. Still encountered in fellowship meetings and informal, after-church conversations, here is one of the all-time classics.

According to this deception, evolution is impossible because it violates the Second Law of Thermodynamics which, we are told, states that entropy, a scientific term for disorder, always increases. In a universe of ever increasing disorder, no organism could naturally develop any greater degree of complexity. Obviously this makes almost every evolutionary improvement fundamentally inconceivable.

But what does the Second Law of Thermodynamics actually say? Rather than looking it up in an elementary physics textbook, which can be found in any public high school, and rather than searching for it among educational web sites offering truthful knowledge of science, let's instead refer to instructive books written by Christians for Christians. Henry Morris, with a Ph.D. in Hydraulic Engineering and one of the founders of the Institute for Creation Research, presented it this way:

> All processes manifest a tendency toward decay and disintegration, with a net increase in what is called the *entropy*, or state of randomness or disorder, of the system. This is called the Second Law of Thermodynamics.[35]

In David A. Noebel's massive Christian compendium *Understanding the Times: The Religious Worldviews of Our Day and the Search for Truth*, the Second Law of Thermodynamics is stated as follows:

> In a system, the amount of disorder, or entropy, must increase and the amount of useful energy must decrease.[36]

These absurd statements say nothing truthful about the physical reality in which we live. All of our children, each one of whom certainly qualifies as a physical system, do not "manifest a tendency toward decay and disintegration." Instead of growing up into an adult, every child would simply wither away and die if it were true that entropy "must increase and the amount of useful energy must decrease." You will not find either of the above statements in any physics textbook, written in any language, anywhere in the universe.

Here is how the Second Law of Thermodynamics is presented in an introductory physics class:

> In an isolated system which is not in equilibrium, entropy will increase until the system reaches equilibrium.[37]

As you can see the Second Law does not say that entropy increases – it says entropy *will* increase when certain *conditions* are met. The term "isolated system" is defined as a system which is not receiving any energy – it is *isolated* from energy. A refrigerator is a device which decreases the amount of entropy within its compartments. It can transform chaotic, high entropy water into organized, low entropy ice, but only if you supply the energy the mechanism needs to do this work. Turn the refrigerator into an isolated system by cutting off the supply of electricity and the rise of chaos within the compartments will ensue. A large metropolis can maintain, and even increase, its internal, highly organized components thanks to a supply of energy. But if the flow of energy comes to a halt, the entire organization begins to deteriorate – the loss of energy begins an increase in entropy.

The entire universe appears to be an isolated system, although this cannot be said definitively because the full nature of the universe is not yet known. The earth, however, is definitely not an isolated system since it constantly receives energy from the sun. Living organisms also are not isolated systems. All living things absorb energy from their environment in one way or another and use it to counteract increases of entropy.

A woman who discovers she is pregnant knows that over the next nine months she will take in a lot of energy and become increasingly complex. A tiny newborn will consume a great deal of energy while growing up into a six-foot tall, socially sophisticated adult of extraordinary complexity. And as a final example, when you cut your finger you don't bleed to death with an ever increasing flow of blood, as would be the case if "entropy always increases." Your body uses its energy reserves to staunch the flow of blood and rebuild the damaged tissue – taking you back to your former, very organized self. All of these well known characteristics of our biological reality are ultimately driven by the energy radiating from the sun.

The Second Law of Thermodynamics does not say that entropy always increases, it does not prevent living organisms from becoming more complex, and it does not prevent evolution. The Christian evolution-deniers who make this preposterous statement try to replace our common understanding of everyday life with a ridiculous strawman.

8

Pulverized by Probability

Among the standard arguments designed to impress the congregation as well as equip anti-evolution apologists for skirmishes with the defenders of evolution, this one rises to almost infinite heights – an awesome giant of a strawman intent on crushing evolution with all the power one can summon from a hammering of genuinely massive numbers. According to this popular beguilement, a simple probability calculation can easily demonstrate that evolution is not merely impossible – it is fantastically impossible.

Living cells contain long strands of DNA molecules which carry the coded information used in regulating the life, the propagation, and the evolution of contemporary organisms. It would not be unusual for one of these strands to contain literally hundreds of millions of molecular codes. Suppose we represent just a tiny fragment of DNA with a sequence of letters and periods only thirty characters long. If we begin with a scrambled sequence like this:

```
TLSXHOEGVJ..CFDLAKKHMFB.AV.SQW
```

and we line these characters up across an extra wide slot machine and randomly change them with each pull of the lever, about how many times would we have to pull that lever and about how long would it take until this meaningful result finally appears?

```
HAVE.YOU.EVER.BEEN.MISINFORMED
```

The total number of different thirty-character sequences made up of capital letters and periods is, approximately, eighty seven followed by forty one zeros.[38] This is a spectacularly large number. The purely random appearance of the above statement is therefore spectacularly unlikely. But exactly how unlikely is it? To run through all of the different thirty-character sequences at a speed of a million every second would take many *billions* of lifetimes of the universe![39] The chances of even a minor DNA sequence randomly appearing as something meaningful to life are therefore stupendously remote. Since evolution is based on random mutations, evolution is clearly an impossibility, even over far more than the lifetime of the universe.

You are sitting at a table. On the table in front of you are two identical computers. One of them is working endlessly on the above problem, generating a new, randomized thirty-character sequence every millionth of a second and then, within that same flick of time, testing the result to see if it equals the organized message we are looking for. We already know it's very likely the universe is going to die of old age many times over before the computer happens upon the correct sequence. The other computer works away on the same problem, also generating one random sequence of thirty characters after another, but it produces the correct result in less than a second.

The second computer cuts through the seemingly infinite number of possibilities because whenever it finds a correct character in the correct position, it stops accepting any more random characters for that position. This efficient solution is a simulation of the way evolution constructs new creations by building on past successes.

Evolution-deniers present a model of a universe whose one and only characteristic is the occurrence of random events. But we don't live in such a simplistic universe. We live in a universe governed by laws of nature. Those laws impose themselves on random events. We don't necessarily need to use an idea from biology to arrive at the evolution-denier's "impossible" result ten to the thirty-six times faster than they do. Instead, we could use our knowledge of, say, chemistry to model groups of elements which commonly appear in nature by increasing the random appearance of well known combinations of letters such as ER, QU, MIS, OU, IN, CK, and ED – to name just a few letter-molecules. Or we could add some physics to the solution by giving the period character a force which repels other periods. This guarantees they will never appear side-by-side, thereby turning the period into a powerful word separator. We could also combine ideas from all three subjects – biology, chemistry and physics, and arrive at the result even faster. The iniquity of evolution-deniers is that they have taken the worst possible solution to this problem and falsely presented it as though it were the only possible solution.

Since we began by working this problem with a biologically inspired solution from the real world, let's go some distance with biological correctness and expose a few more realities which evolution-deniers have deliberately hidden from view.

Never mentioned in the plodding computational grind these deceivers sell to their fellow Christians is the most obvious fact of all – that there would never be just one character string mutating all by itself to the end of the universe. Instead, many copies would be distributed throughout a population which would greatly increase the chances of a useful mutation. Some populations of organisms are astoundingly large, such as bacteria, which have been measured at a million per gram of topsoil.[40] Look up the title of the reference in the back, do the math and you will discover that one ton of topsoil from the Omo Biosphere Reserve in southwestern Nigeria contains, approximately, one trillion of these micro-beasts. There are a lot of tons of topsoil on the earth, and that's just the topsoil. There are between one million and fifteen million bacteria in a teaspoon of sea water.[41] There is also a gigantic bacterial biosphere under our feet extending down through kilometers of rock.[42] In every one of these bacteria there are long strands of DNA to represent many, many segments equivalent to the thirty-character string. Any acknowledgment of the massive size of the earth's biomass always seems to be conveniently missing from the evolution-deniers' perfidious presentations.

Nor is it ever mentioned that all of this random mutating does not have to produce a perfect result. There can be incorrect characters (misspellings) in the sequence and it will still work just fine biologically. We know this to be true since it is a well known fact that most mutations have little or no effect on living organisms. Currently there are more than a thousand known mutations of human hemoglobin and about one person out of every two thousand is living with one of them.[43] While a few of these mutations cause devastating medical problems, most have no effect. The billions of people on earth do not all need a perfectly "spelled" hemoglobin molecule. In fact, since there is such a variety of different yet workable hemoglobin molecules, the perfect such molecule cannot even be defined. This tolerance for imperfection further reduces the seemingly infinite workload.

Also deliberately excluded from the presentations by evolution-deniers is any acknowledgment of the many other meaningful phrases which can also be stated in only thirty characters, such as this one:

YYOU..ARE...JQ.INFORMED.D.GCAX

Even in a small sequence of DNA there may be many biologically useful molecules, not just one as implied by the not-so-informative designers of the strawman. It seems the strawman was crafted to be as complete a fake as possible.

Fervently constructing a world view assembled out of lies about science, evolution-deniers have absolutely no problem designing misinformation for their fellow Christians. As Christians themselves they expect a lot of latitude and they get it. In group fellowship talks and in evening video presentations on anti-evolution and intelligent design, the members of the congregation sit in the church occasionally remarking to one another, "Isn't that fantastic!" As a matter of fact it is much more than just fantastic – it is mythological.

9

From the Library of False Teachings

In accumulating their collection of fraudulent arguments, the most common target for the craftsmen of evolution-denial is, of course, the theory of evolution itself. By providing false descriptions of the theory of evolution, false teachers conjure up the illusion of "problems" with evolution and seek to portray evolutionary scientists as corrupt, or imbecilic, or both. Here is a recent and remarkably blatant example, granted immortality in print and scattered throughout the Christian educational system.

Introductory biology textbooks in high schools, colleges, and universities generally strive for a full-blown multimedia experience. It sometimes can be difficult finding a page displaying nothing but text. But in 1859 when Mr. John Murray published Charles Darwin's seminal work, only one illustration was included in *The Origin of Species*, all the rest was hundreds of pages of text. The drawing doesn't even have a page number – it's inserted in the first and second editions between pages 116 and 117, and is referred to throughout the book as "the diagram."[44]

If you were Darwin and you'd been working on a five hundred page manuscript for more than twenty years, covering a vast subject ranging widely across the highly visual and diverse sciences of biology and geology, what would you pick as the single most important illustration in all of evolution? Perhaps you would choose a drawing of different species of finches inhabiting the several Galapagos Islands, displaying the variations of their beaks and thereby demonstrating the long-term effects of geographical isolation. Or perhaps you would portray an eroded hillside exposing geologic strata with imbedded fossils, displaying the spread of "organic beings" into new environments – from sea to land to air – over the long passage of the ages. Or perhaps you would provide a drawing of a series of skeletons of different animals, demonstrating the underlying structural relatedness of those assorted species. For his one and only artistic composition, however, Darwin settled on a figure far more instructive than any of these. The most important illustration in all of evolution is, from Darwin's own hand, a simple line drawing depicting evolutionary descent.

As in the genealogical descent of your own family, showing as many of your relations as you can account for, evolutionary descent is a wide branching structure, but on a much greater time scale. And although your personal genealogy is a schematic history of past and present individuals, Darwin's illustration of "descent from a common ancestor" is a history of past and present populations. For example, a small extract from the modern version of "the diagram,"

Earlier Primates

Common Ancestor

Chimpanzees   Human Beings

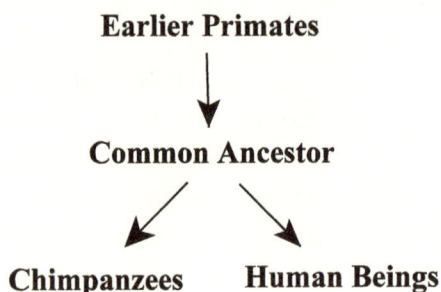

Figure 1: Branching Descent

displaying a simplified view of humanity's great epic, is shown in Figure 1. As you can see, evolutionary theory does not say that humans evolved *from* chimpanzees, as would be portrayed in a linear diagram such as Figure 2. The sequential descent displayed in Figure 2 is a false depiction of evolution. Instead, the theory of evolution describes humans and chimpanzees evolving simultaneously from a common ancestor, with perhaps both sides changing quite a bit over millions of years.

As already mentioned, an evolutionary "common ancestor" is never an individual. There are no individuals in evolution, only populations. Human evolution, for example, may have arisen when a group of jungle primates were forced by competition with others of their species to inhabit a large valley at the edge of the jungle. As the climate slowly changed, the jungle lost ground to the advance of a sparse forest. But the group retained their territory in their bountiful valley while other related groups moved away with the retreating jungle. This isolated valley group, with its pool of

Earlier Primates

↓

Chimpanzees

↓

Human Beings

Figure 2: Sequential Descent

genes slowly changing over hundreds of generations, became one of our many common ancestors.

The branches of evolutionary descent account for the extravaganza of diversity we see in biology today and at any time in the fossil record. We find in Africa, for example, monkeys, chimpanzees, and humans, each displaying various degrees of difference, but all living together in the same landscape with each group avoiding competition with the others by inhabiting their own more or less independent ecological niche. In this same landscape we also find other mammals along with reptiles, amphibians, and fish – all related by various common ancestors from different periods in the distant past. Evolutionary descent even accommodates a loss of features, such as the wayward evolution of blind fish isolated in permanently dark caves.[45] Like an ancient, robust tree with both live and dead branches, Darwin's simple diagram of common descent has enormous explanatory power and so it is, as he well knew, the single most significant illustration in all of evolution.[46]

Lobe-finned Fish

↓

*many extinctions
many new species*

↓

Common Ancestor

↙ ↘

Lobe-finned Fish   **Fish-Amphibian
Transitionals**

↓

*many extinctions
many new species*

*many extinctions
many new species*

↓

↓   **Common Ancestor**

↙ ↘

Lobe-finned Fish   **Fish-Amphibian
Transitionals**   **Amphibians**

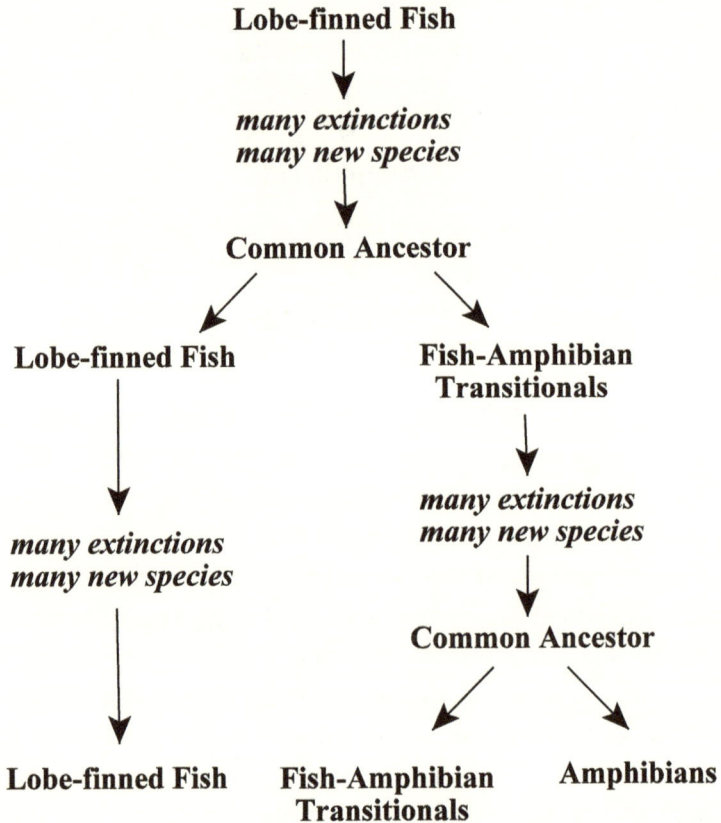

Figure 3: Evolution of Amphibians

Another example of the branching structure of common descent, in this case as it applies to the evolution of amphibians from fresh water lobe-finned fish between 380 and 360 million years ago, is presented in Figure 3. This diagram depicts a large but unknown number of species, only a few of which have been preserved in the fossil record, and only a few of those have so far been discovered. As Darwin explained:

> We see nothing of these slow changes in progress, until the hand of time has marked the long lapse of ages, and then so imperfect is our view into long past geological ages, that we only see that the forms of life are now different from what they formerly were.[47]

Based on how the extinctions and new species play out in Figure 3, at any particular moment in time there could be quite a varied mix of creatures all exhibiting different degrees of advancement, just as we see an enormous variety in biology today. By the time the amphibians begin to appear in the fossil record, they're likely to be accompanied by fossils of lobe-finned fish as well as fossils of various fish-amphibian transitional animals. There is every reason to expect that examples of all these different "organic beings" would be living together in the same environment but avoiding competition with one another by occupying slightly different ecological niches.

Nobody could honestly mistake Figure 3 as a simple linear sequence of improvements just as nobody could honestly mistake Darwin's rather large and intricate original drawing, along with his lengthy and meticulous description of it, as a portrayal of linear, or sequential, descent. But honesty is always a choice for people to make. Here is an example of crafted misinformation from the Christian organization Reasons To Believe. In an article disparaging the theory of evolution and its explanation for the origin of amphibians, Fazale R. Rana, with a Ph.D. in Chemistry, the

Executive Vice President of Research and Apologetics at Reasons To Believe, and a lecturer for the Master of Science and Religion program in Christian Apologetics at Biola University, explains the supposed wrong-headedness of evolutionary scientists:

> Yet another concern for the evolutionary account is that the fishapods co-occur in the fossil record. Instead of appearing in a sequential fashion, the creatures evidently coexisted and overlapped. In other words, the pattern observed in the fossil record doesn't describe a linear evolutionary change over time. The challenge for evolutionary biologists lies in letting the fossil record dictate the pattern rather than imposing a pattern on the data.[48]

Reasons To Believe wants you to believe that evolutionary scientists claim the fossil record *should* reveal "linear evolutionary change." That's the pattern Dr. Rana implies these scientists are trying to impose on the data. Since the untidy assortment of "fishapod" fossils does not display a series of advancements "appearing in a sequential fashion," you are supposed to conclude that evolutionary scientists, for reasons one can only imagine, have been trying to sell the world a bogus theory for the last one hundred and fifty years. Dr. Rana would also like you to believe that generations of biologists have spent the last one hundred and fifty years ignoring the plain facts of the fossil record as well as the plain language of Charles Darwin. But the truth, of course, is just the opposite. All that overlap Dr. Rana writes about is precisely what evolutionary scientists expect to see from the simple branching structure of common descent, exactly as Darwin so carefully described with his single, famous diagram.

Reasons To Believe has published in timeless print this false description of the theory of evolution in order to make their own contribution to a false depiction of evolutionary scientists. Especially notable is the fact that Dr. Rana's concoction avoids saying straightforwardly that biologists claim the fossil record should demonstrate linear descent. That idea is slipped into the mind sideways as the reader naturally deduces it from the phrasing of the statements. Ideas and conclusions presented via insinuation and rhetorical trickery are commonplace in the works of Christian evolution-deniers. In defending Christianity, a use of language that's plain and helpful, honest, open and free of guile, would force the removal of every brick from the wall of anti-evolution.

If you also wish to become a false teacher and make your own contributions to a voluminous library of lies, consider getting a Ph.D. first – not the better to inform yourself or others, but the better to disguise yourself. And to tell a lie so obvious as evolutionary scientists expecting to see "linear evolutionary change" in the face of Darwin's one and only diagram, be sure to throw into your equipage a hefty collection of denial, arrogance, and contempt for the intelligence of ordinary people, convince yourself that if it isn't corruption which rules the hearts of evolutionary scientists, then certainly imbecility must govern their minds, nurture the conviction that any accusations of dishonesty delivered upon you are patently false simply because your work is for the Lord, and believe in all the rewards which await you in heaven not only for your good-natured sacrifices and hard work, not only for your sincere faith, but also because you see that you are a warrior in the war between good and evil and there are necessary burdens you must bear under the auspices of God, including, it seems, the publication and distribution of works of overt deceit.

Just as the Catholic Church created an ignominious and immortal icon with its abuse of Galileo, including the four centuries it took to apologize for it, so also will this construction of a library of lies stand through time as a constant reminder of the corruptions of a corrupt people. The irony, of course, is that it is in fact a library of the truth.

10

Mining and Refining

In the theater of evolution-denial, where half-truths perform as though they were whole-truths, where what is clearly wrong joins the tragedy claiming it is right, and where it always seems as though the author of the script has no angel for a muse – on this dishonest stage the words of scientists become just another instrument in a orchestration of deceit. The late Dr. Colin Patterson, senior paleontologist at the British Museum of Natural History, has been quoted as saying the following:

> I will lay it on the line – there is not one such [transitional] fossil for which one could make a watertight argument.

When a Hollywood movie achieves great success, playing in theaters on a seemingly endless run, they say the movie has "legs." Colin Patterson's quote has "legs." It is scattered across the landscape of evolution-denial on a seemingly endless run. You can search for it fairly easily and find it employed at a long list of Christian web sites.[49] It has been immortalized in print in such Christian books as *The Creation Hypothesis* by J. P. Moreland (Editor), and David A. Noebel's *Understanding the Times*, and many others.[50] It can be found in anti-evolution videos on YouTube.[51] You can also see it in "The Truth Project," a DVD series from the popular Christian organization Focus On The Family.[52]

The quote appears to reveal a prominent evolutionary scientist expressing some doubt concerning the veracity of the theory of evolution. To be specific, the statement is presented by evolution-deniers as an example of a paleontologist, and not just any paleontologist, but one who presumably has access to almost all the fossils of the world, disclosing that there are no valid transitional fossils demonstrating the evolution of one type of animal from another. But is this really what Dr. Patterson was saying?

The quote is not complete. A complete quotation would have to include enough information to provide the context so that the true meaning of the statement can be conveyed. Christians know all about the importance of proper context. They often remind each other not take scripture out of context. But proper context is always an intolerable requirement to anyone intent on deforming the truth.

A typical transitional fossil would be the famous Lucy (*Australopithecus afarensis*), which exhibits numerous pieces of evidence for the transition of humans from more primitive primates.[53] The transition of birds from small theropod dinosaurs can be seen in the spectacular *Archaeopteryx*.[54] The equally fabulous *Tiktaalik roseae* attests to the transition of amphibians from freshwater lobe-finned fish – limbs from fins.[55] But as it turns out Dr. Patterson was not talking about just any transitional fossil. Instead, he was specifically

referring to that most extraordinary and improbable transitional fossil of all – one which is the direct, individual
ancestor of a species, and as he stated:

> I will lay it on the line – there is not one such
> [transitional] fossil for which one could make
> a watertight argument.

Other than the Emperor of Japan, no one in the world
today can trace their ancestry back to some individual two
thousand years ago.[56] How then could a paleontologist
possibly trace the ancestry of some animal back to an
individual fossil that is millions of years old? In fact anything
even remotely resembling a lineage of individuals will never
appear in the fossil record because the rarity of fossils makes
it abundantly clear that almost all "organic beings" never get
fossilized, most that do are hidden inside solid rock – in most
cases thousands of feet of solid rock, and those fossils which
do naturally become visible are in the process of being
destroyed by erosion.

Criticizing paleontologists for their inability to acquire
impossible evidence might be valid if it weren't for the fact
that they don't need any direct-in-line ancestor fossils in
order to verify the evolutionary history of life. The evolution
of new species arises from many mutations shared within an
entire population and not from identifiable, individual
ancestors. The rare fossilization of individuals simply creates
fortuitous snapshots of some of these evolving populations.

If the population of an animal were geographically
divided, for example by rising ocean levels isolating a large
group on a newly formed island, or simply by dispersing the
population over a large, varied landscape broken up by

mountains, valleys, rivers and deserts, then across a span of time measured in many, many generations, the separated groups would experience different sets of random mutations, thereby creating distinct sets of genetic codes. However, just because some of the DNA codes are different doesn't mean that the separated groups have become unique species. On rare occasions in the wild, grizzly bears and polar bears do sometimes mate and the result is a fertile offspring.[57] Also there are obviously many genetic differences among human beings, yet we can still form successful families anywhere on earth. But eventually the genetic differences between separated groups would accumulate to such an extent that their genetic material could no longer be shared, at which point it no longer matters if they remain physically divided since now they are genetically isolated from each other. They have become different species.

There are many examples in biology of the genesis of new species.[58] Some of the most interesting and revealing are called "ring species." They form a ring of subspecies around mountain ranges or large valleys. The lungless salamander *Ensatina* has seven distinctly colored subspecies whose habitats ring California's central valley like the numbers on a clock.[59] All the adjacent subspecies can interbreed successfully except for one pair of neighbors at the southern end of the valley. Over the many generations it took the salamander to migrate around the huge valley, so many genetic differences accumulated along the way that by the time the long journey was complete, the initial subspecies and the final subspecies could no longer interbreed when they finally met in the south. Nobody needs to accomplish the impossible task of tracing all of the direct ancestors of all these salamanders to prove that the initial and final subspecies are related to each other since the interbreeding ability of all the neighboring subspecies clearly establishes their relatedness.

While ring species demonstrate the gradual development of a new species via minor evolutionary changes over relatively short periods of time, transitional fossils demonstrate major evolutionary changes which accumulate over much longer periods of time. But here again scientists don't need any direct-in-line ancestor fossils in order to demonstrate that one group of "organic beings" provided the foundation for the evolutionary development of another group. There are plenty of other non-ancestral fossils which accomplish that.

The fossil Lucy, for example, may or may not be a direct ancestor of ours. She may have died before having any offspring, or she may have died along with all of her offspring. Far more likely, however, she may have been part of a population which branched off from the human line and went extinct millions of years ago. But as scientists have repeatedly pointed out, if there was no evolution of humans from more primitive primates, there would be no humanoid fossil Lucy, nor any of the many other humanoid fossils which have been found. If there was no evolution of birds from dinosaurs there would be no fossil *Archaeopteryx*, nor any of the many other dinosaur-bird transitional fossils which have been found. If there was no evolution of whales from four-footed land animals there would be no fossil *Ambulocetans natans*, nor any of the other remarkable fossils demonstrating the transition of the cetaceans from land to sea.[60] All of these impressive transitional fossils, and many others like them, are easily explained by the evolutionary history of life and their existence is predicted by, and thereby validates, the theory of evolution (see Chapter 15).

Colin Patterson's quote is not an example of a famous scientist expressing doubts about the theory of evolution. He is not saying that there are no transitional fossils, nor is he implying that the theory of evolution is without evidence, merit, or authenticity. Instead, Dr. Patterson is saying that there are no fossils which have been, or even can be, identified as individual, direct ancestors of a species. As he said in his book *Evolution*:

> Fossils may tell us many things, but one thing they can never disclose is whether they were ancestors of anything else.[61]

– a statement he repeated when he said:

> I will lay it on the line – there is not one such [transitional] fossil for which one could make a watertight argument.

But just to be *absolutely* certain of the meaning of this quote, Lionel Theunissen inquired of Dr. Patterson himself, suggesting that he, Dr. Patterson, was referring to the mythical and irrelevant direct ancestor fossils, rather than to transitional fossils in general, and received a reply which included the following:

> The specific quote you mention, from a letter to Sunderland dated 10th April 1979, is accurate as far as it goes. The passage quoted continues " . . . a watertight argument. The reason is that statements about ancestry and descent are not applicable in the fossil record.

Is *Archaeopteryx* the ancestor of all birds? Perhaps yes, perhaps no: there is no way of answering the question. It is easy enough to make up stories of how one form gave rise to another, and to find reasons why the stages should be favoured by natural selection. But such stories are not part of science, for there is no way to put them to the test."

I think the continuation of the passage shows clearly that your interpretation (at the end of your letter) is correct, and the creationists' is false.[62]

When Mr. Theunissen presented this clarification to the Creation Science Foundation, he received a lengthy and dismissive reply from its director, Carl Wieland, which included these comments:

I do NOT agree that the continuation shows clearly that your interpretation is correct. Nor is it fair for Patterson to comment on the creationist interpretation without a clearer definition of what is meant by 'transitional forms' . . .

Not fair for Dr. Patterson to comment on someone's misinterpretation of his own words? Would it also not be fair for the Apostle Paul to make any corrective comments on someone's misinterpretation of his own words? Is it also not fair for you to comment on someone's misinterpretation of what you have said? And is Dr. Patterson's final statement, "your interpretation . . . is correct, and the creationists' is false" somehow lacking sufficient clarity?

As if this were not a perfect enough example of deliberate misrepresentation by Christian evolution-deniers, it gets worse. Shortly after these exchanges Mr. Theunissen discovered that an out-of-context excerpt from the letter he had just received from Dr. Patterson was being posted around the Internet claiming that the creationist misinterpretation was actually valid. Here is the text which the creationists extracted:

> The specific quote you mention, from a letter
> to Sunderland dated 10th April 1979, is accu-
> rate as far as it goes.

Anyone with an immoral intent can go through this book and selectively pull out text to falsely show doubts about evolution and falsely show support for evolution-deniers. Who would expect anything less of the wicked?[63] In the culture of evolution-denial, even attempts at clarification are assiduously twisted into misinformation. By sifting through the literature and dialogues of scientists, everyone from Charles Darwin to Carl Sagan and dozens of others you've probably never heard of, the dedicated quote miners of evolution-denial dig out the nuggets, each one as phony as a three-dollar bill. Having extracted the counterfeit ore, they even refine it, as they did by isolating and publicizing an easily misinterpreted sentence from Patterson's reply to Theunissen.

In defining hypocrisy there is hardly a better example than the Christian concern for accurate scriptural context and the abuse of scientific context by Christian evolution-deniers.

No matter how often they are exposed and explained, the deliberate misrepresentations of statements by scientists keep coming back. It seems hypocrisy also has the power to resurrect. This is what Christian evolution-deniers bring with them to Sunday school.

11

Insulted in Every Possible Way

Just before sunrise the men began to assemble at the church. They're all generous, family men, each one having dedicated himself decades ago to the Christian faith. All but a few are already years into their retirement. Who else would get up at this hour for a church event?

Their knowledge of scripture is impressive. Over the years most of them have attended numerous Bible-study classes, some have even run such classes, or are currently doing so. Certainly almost all of them go over passages of scripture everyday, including a few who lay out a daily allotment to accomplish yet another annual reading through the entire Bible.

This morning's event is principally a social gathering. It's a men-only fellowship breakfast which takes place twice each month and it's not to be missed – the pastor's wife prepares the course and she's a fabulous cook. Much less of a draw, but definitely worth the meal, is this morning's after-breakfast talk on the history of medicine delivered by a member of the group, a retired surgeon.

The breakfast, as usual, fulfilled its reputation, especially the ladle full of blueberries in syrup. The MC primed the group with a few good jokes and then the presentation began.

This informal journey through medical history went along an interesting and informative track beginning in the Middle Ages, then on through the accomplishments of the Renaissance, followed by a rising level of achievement during the Enlightenment, until 1859 when, we were told, Darwin published *The Origin of Species*. It was then mentioned emphatically, twice, that Charles Darwin was a "vicious racist." Following this discordant aside, the beat was picked back up and the talk continued on into increasingly more modern facts.

Speaking of facts, here are some facts from a traveler spending some of his time in what he referred to as "a slave country:"

> I may mention one very trifling anecdote, which at the time struck me more forcibly than any story of cruelty. I was crossing a ferry with a negro, who was uncommonly stupid. In endeavouring to make him understand, I talked loud, and made signs, in doing which I passed my hand near his face. He, I suppose, thought I was in a passion, and was going to strike him ; for instantly, with a frightened look and half-shut eyes, he dropped his hands. I shall never forget my

feelings of surprise, disgust, and shame, at seeing a great powerful man afraid even to ward off a blow, directed, as he thought, at his face. This man had been trained to a degradation lower than the slavery of the most helpless animal.

Charles Darwin
H. M. S. Beagle
April 1832, Brazil[84]

12

BS

There are always controversies in science. It is one of the profession's most exciting and challenging characteristics. Recently there has been a controversy regarding the demise of the dinosaurs. Many scientists are convinced that an asteroid or comet smacked the earth and wiped them out. But some contend this explanation because the extinction, they claim, does not coincide with the big blast from space. This debate can be found in peer-reviewed scientific journals where the geologists and paleontologists involved present the facts they have gathered by doing the hard work of scouring the world's landscapes and oceans for convincing evidence.[64]

Peer-review is a long accepted method of maintaining the standards of excellence in reporting scientific evidence, including contentious evidence which may stand against widely accepted ideas, as in the above example. The peer-review process works as a filter, capturing and discarding the writings of incompetents, charlatans, and crackpots. Some false or even deceitful information which does happen to get published now and then is weeded out later when others fail to observe the claims.[65] In the imperfect world of human nature, science is thereby self-correcting.

It has been said by many proponents of intelligent design that they don't need to publish their "scientific" ideas and any supporting evidence in peer-reviewed scientific journals. They claim that it's sufficient to present their assertions by other means, such as in the books they write. After all, they say, that's exactly what Darwin did – he presented the theory of evolution to the scientific community by writing a book about it. But Charles Darwin actually did much more than just publish a book about his ideas. To physically validate as many of those ideas as possible, Darwin included in his book numerous results from his extraordinarily wide-ranging experimental research.

Most scientific papers begin with a brief abstract summarizing the contents therein. But in his Introduction to *The Origin of Species*, Darwin referred to the entire work as "this Abstract."[66] At five hundred pages it is perhaps the longest abstract in the history of science. He was inclined to this definition because, in rushing the book into print, he was forced to leave out all of the notes and references he had intended to include. Those were promised to appear in a later edition after the two or three years he estimated it would take to assemble them all. Nevertheless, in this long "Abstract" there are mentioned numerous, repeatable experiments and other systematic observations which Darwin personally conducted, demonstrating his professional commitment to confirming the reality of his ideas. Here in his own words are three examples of the scientist at work:

From Chapter III, "Struggle for Existence:"

> With plants there is a vast destruction of seeds,
> but, from some observations which I have
> made, I believe that it is the seedlings which
> suffer most from germinating in ground al-
> ready thickly stocked with other plants. Seed-
> lings, also, are destroyed in vast numbers by
> various enemies; for instance, on a piece of
> ground three feet long and two wide, dug and
> cleared, and where there could be no choking
> from other plants, I marked all the seedlings of
> our native weeds as they came up, and out of
> the 357 no less than 295 were destroyed,
> chiefly by slugs and insects. If turf which has
> long been mown, and the case would be the
> same with turf closely browsed by quadru-
> peds, be let to grow, the more vigorous plants
> gradually kill the less vigorous, though fully
> grown, plants: thus out of twenty species
> growing on a little plot of turf (three feet by
> four) nine species perished from the other
> species being allowed to grow up freely.[67]

From Chapter IV, "Natural Selection," Darwin investigated
a plant's control over bees to facilitate pollination:

> Some holly-trees bear only male flowers,
> which have four stamens producing rather a
> small quantity of pollen, and a rudimentary
> pistil; other holly-trees bear only female flow-
> ers; these have a full-sized pistil, and four

stamens with shrivelled anthers, in which not a grain of pollen can be detected. Having found a female tree exactly sixty yards from a male tree, I put the stigmas of twenty flowers, taken from different branches, under the microscope, and on all, without exception, there were pollen-grains, and on some a profusion of pollen. As the wind had set for several days from the female to the male tree, the pollen could not thus have been carried. The weather had been cold and boisterous, and therefore not favourable to bees, nevertheless every female flower which I examined had been effectually fertilised by the bees, accidentally dusted with pollen, having flown from tree to tree in search of nectar.[68]

From Chapter XI, "Geographical Distribution:"

Freshwater fish, I find, eat seeds of many land and water plants: fish are frequently devoured by birds, and thus the seeds might be transported from place to place. I forced many kinds of seeds into the stomachs of dead fish, and then gave their bodies to fishing-eagles, storks, and pelicans; these birds after an interval of many hours, either rejected the seeds in pellets or passed them in their excrement; and several of these seeds retained their power of germination.[69]

Many more examples of Darwin's experimental work can be

found throughout *The Origin of Species*, including breeding experiments with pigeons, a microscopic examination of the interactions between ants and both adult and juvenile aphides, the ecological effects of establishing a new forest of Scottish firs on heath land in Staffordshire, the fertilization of the common red clover *Trifolium pratense* by "humble-bees" only, and the survivability of various seeds immersed for various lengths of time in sea water.[70] In contrast to this wealth of hands-on science from Charles Darwin, what do we have from the professional scientists promoting intelligent design?

One of the most revealing aspects of the intelligent design "challenge" to evolution resides in the complete lack of experimental activities by the very few research scientists supporting these ideas. While they publish books and numerous articles for Christian newsletters, speak before groups, make appearances in videos, hammer away on the Internet, and even offer testimony in court under oath, they never actually engage the scientific process, though it is their own profession. Not only do they not offer papers on the topics of intelligent design and irreducible complexity to peer-reviewed scientific journals, more importantly they don't even do any of the groundwork of experimental research on these topics, including experiments they themselves repeatedly suggest.

Nine years is a long time in our frenetically competitive scientific culture. Consider how much the gadgetry of technology has changed over just the last nine years. Nine years after publishing *Darwin's Black Box*, Dr. Michael Behe, who is a tenured research biochemist at Lehigh University and has published research papers on topics other than intelligent design and irreducible complexity in peer-reviewed scientific journals, gave the following testimony in the federal court case on the teaching of intelligent design in science classrooms in the Dover, Pennsylvania public schools:[71]

> [Note: The "Plaintiffs" in this case were opposed to the teaching of intelligent design as a science.]

> For the Plaintiffs, MR. ERIC ROTHSCHILD: Okay. So in terms of irreducibly complex structures, you haven't done any tests, right?

> DR. MICHAEL BEHE: That's right.

> MR. ROTHSCHILD: You're not planning on any tests—

> DR. BEHE: That's right.

> MR. ROTHSCHILD: —of the type you described here?

> DR. BEHE: Well, I'm doing my theoretical work with David Snoke and hope to continue that, so I think that bears on this question.

> MR. ROTHSCILD: Bears on it, but it's not testing an irreducibly complex system in the way you described in this article?

> DR. BEHE: That's right.

MR. ROTHSCHILD: And nobody else, you're not aware of anybody else in the intelligent design movement doing a test of the type you described here of an irreducibly complex system?

DR. BEHE: No, not yet.

Not yet? Several days later Dr. Scott Minnich, a research microbiologist at the University of Idaho who, like Dr. Behe, has also published research papers on topics other than intelligent design in peer-reviewed scientific journals, including research relevant to the proposed "test" being discussed in this court case, gave similar testimony:[72]

For the Defendants, MR. ROBERT MUISE: Dr. Minnich, another complaint that's often brought up, and plaintiffs' experts brought it up in this case, is that intelligent design is not testable. It's not falsifiable. Would you agree with that claim?

DR. SCOTT MINNICH: No, I don't. I have a quote from Mike Behe. "In fact, intelligent design is open to direct experimental rebuttal. To falsify such a claim a scientist could go into the laboratory, place a bacterial species lacking a flagellum under some selective pressure, for motility say, grow it for ten thousand generations and see if a flagellum or any equally complex system was pro-duced. If that happened my claims would be neatly disproven."

MR. MUISE: Is this an experiment that could be done in a lab?

DR. MINNICH: It could be, and I, you know, would say that, you know, up the ante. I'll give somebody a type three secretory system intact and the missing proteins required to convert it into a flagellum and let them go, see if you can get a flagellum from a type three system. That's a falsifiable doable experiment. That's just the type of experiment that could be subjected to this type of analysis.

MR. MUISE: Would this be an experiment that you would do?

DR. MINNICH: You know, I think about it, I would be intrigued to do it. Knowing the tolerance limits for these proteins and how they would assemble I wouldn't expect it to work. But that's my bias.

MR. MUISE: You think natural selection could account for that, take the type three secretory system, the additional proteins, and see if natural selection can build a bacterial flagellum from that?

DR. MINNICH: I'm not convinced that it

could, but again it's a plausible experiment.
They should write a grant and see if we can
do it.

They should write a grant? During cross-examination the
next day, the topic was brought up again:[73]

> For the Plaintiffs, MR. STEPHEN HARVEY:
> Now you claim that intelligent design can be
> tested, correct?

> DR. SCOTT MINNICH: Correct.

> MR. HARVEY: Matt, please bring up slide 40.
> And that's your claim right there that you put
> up during your direct testimony to state that
> intelligent design can be tested, right?

> DR. MINNICH: Right. I think it's falsifiable.

> MR. HARVEY: And neither you nor Dr. Behe
> have run that test, have you?

> DR. MINNICH: We talked about that yester-
> day. And I even, I think, gave a – an experi-
> ment that would be doable. And in thinking
> about it last night, I might try it to see if I can
> get a type III system to change into a flagel-
> lum.

> MR. HARVEY: You haven't run that test,
> right?

DR. MINNICH: I've done parts of it. I know that the type III secretory system will secrete flagellum.

MR. HARVEY: True or false, you haven't done that test?

DR. MINNICH: No.

MR. HARVEY: Correct? You haven't done that test?

DR. MINNICH: What's the point? I mean—

MR. HARVEY: I'm asking you whether you have done the test that you propose for intelligent design? That's a yes or no question.

DR. MINNICH: No, I have not.

MR. HARVEY: Okay. Now Dr. Behe hasn't either, has he?

DR. MINNICH: I'm not aware of it, no.

There are so many things wrong with this under-oath testimony, it's hard to know where to begin. Perhaps most importantly, the "test" referred to by Dr. Behe and Dr. Minnich – an experiment to see if a bacteria without a flagellum can develop such a locomotive feature – is not a test of intelligent design, as they maintain, but a test of evolution, and a pretty lousy one at that.

Nobody is making the outlandish claim that during the run of this test the flagellum will suddenly appear. If such a miracle were to occur, and occur every time someone repeated this test, it would do far more than show that evolution is false – it would also sweep away everything the Bible says about the need for faith (see Chapter 18).

But if the bacteria develops a flagellum while the scientists observe its step-by-step molecular creation, then evolution is validated and the results become just another entry in a very long list of scientific observations which already verify the theory of evolution (see Chapter 15).

On the other hand, if no flagellum is produced during the run of the experiment, as Dr. Minnich predicts, intelligent design is *not* validated, nor is evolution shown to be false.

A failure of this test does not serve to miraculously erase the massive amount of other, completely independent evidence verifying the theory of evolution (again, see Chapter 15). Instead, in the face of all that other evidence, a test failure would immediately raise obvious questions as to why such a failure would be reasonably expected. Was ten thousand generations enough? Was the mutation rate too low, or too high, or should it have been varied in some way during the test? Was the wrong type of bacteria used? Did the test fail to provide a strong enough selection pressure for motility? Should the selection pressure have varied in some way during the test?

To assume, however, that a failure of this test can provide evidence for intelligent design would make the ridiculous claim that the Designer is observed by the fact that He did nothing. There is no reasonable place on earth where you could get away with a deduction like that, although there are some unreasonable places available. We will, like the lawyers and the witnesses, overlook all the flaws in the purpose of this "test" and just move on.

It is not clear from the testimony whether or not any of the lawyers were aware of the full extent by which they were being deceived. The lawyers for the plaintiffs, who were presenting the case against the inclusion of intelligent design in high school science classes, appear to be content simply to get on record the fact that neither of these two working research scientists had ever done an experiment to backup their statements – not even the test they themselves have repeatedly proposed, even though such work is the daily nature of their profession. That testimony alone is a powerful fact discrediting these evolution-deniers. However, added to this failure to perform is the fact that the proposed flagellum experiment is an easily discernable ruse, at least to anyone in the scientific community. If someone made the claim that man will never fly, nobody has to fly across the Atlantic to prove the statement false. A lesser flight will do.

Experimental research scientists like Dr. Behe and Dr. Minnich, and many others, are generally renowned for their cleverness in designing simple, efficient, and convincing experiments to settle issues. Evolutionary scientists usually have better things to do with their time and money than to give up several years of their careers for each run of this problematical flagellum experiment.[74] After all, they already have an eminently workable theory supported by an ava-lanche of other evidence and can easily explain the outcome of the experiment, no matter what that outcome might be. It's up to the scientists who deny the validity of evolution to design and perform an unambiguous test and then publish the results of their repeatable experiment in a peer-reviewed scientific journal.

The handful of evolution-denying scientists have done everything possible to avoid this work. They have their years of excuses. They are the thinkers – let others do the labor, as in Dr. Minnich's revealing choice of words from his testimony, *"They* should write a grant and see if *we* can do it." Really? *They* should write a grant, Dr. Minnich, and not you? Yet these deniers have never bothered employing their cleverness to come up with an efficient, defining test, although Dr. Minnich was, apparently, "thinking about it last night."

Under the wide spectrum of all the other deceit radiating from the Christian evolution-denial movement, it makes perfect sense for these two working research scientists, and others from their group, to intentionally avoid doing any experiment to support their ideas. Doing so risks exposing the truth they have committed themselves to denying. Despite their I-swear-under-oath-to-tell-the-truth statement that intelligent design can be tested scientifically, it cannot. Science tests nature – there is no test for the supernatural.

A driven misrepresentation of facts and ideas with language comprised of lies, foolishness, deceit, nonsense, exaggeration, insinuation, worthlessness and insincerity is defined in many dictionaries under the word "bull," or more pointedly – "bullshit." Since this word rises from a gut-level disgust, it is sometimes abbreviated in print and speech as "BS" in a attempt to mollify blatant obscenity. In regard to Christian evolution-denial, all the evidence indicates that the word is a perfect fit.

13

What is Life?

In anti-evolution and intelligent design expositions, it's common to link deprecations of evolution with criticisms of biologists for not being able to explain the origin of life. The theory of evolution, of course, has nothing to do with the origin of life – evolution is what happens *after* life first appears. Since science currently offers no explanation for life's beginning, there are those who feel that their non-scientific account somehow adds validity to all sorts of counter-scientific explanations for everything from biology to cosmology. But how believable is it that science will never discover an answer to this fundamental question? According to the advocates of intelligent design, a scientific theory explaining the creation of life will never be achieved for the simplest reason – complexity.

In the popular intelligent design video *Unlocking the Mysteries of Life,* an assumption is made that the very first living thing to appear on this earth had most, if not all, of the features found in a contemporary living cell, including massive amounts of information encoded in DNA along with all of the elaborate molecular mechanisms for transforming that information into the many necessary products of the cell.[75] Apparently ignoring billions of years of cellular evolution is easy if you want to, and with this present-day cell in mind, one of the principle props of evolution-denial, and the central statement of intelligent design, is that life is so phenomenally complex only an entity with the abilities of God could have created it. This claim, however, fails to answer the single most important question in any explanation for the origin of life – what exactly is life?

No one has the answer to this question. If you do, then congratulations, you have acquired an honored place in history. Science has a definition for the seemingly mundane trajectory of an arrow – it's an elliptic path approximating a parabola, thank you Isaac Newton. Science has a definition for gravity – it's a geometric distortion in a space-time continuum, thank you Albert Einstein. Science has a definition for light – it's electric and magnetic waves which are perpendicular to each other and to their direction of propagation, thank you James Clerk Maxwell. But science has no definition for life.

Since all living things today seem to replicate in one way or another, we could try defining life as anything which replicates itself. Unfortunately this simple definition fails because there are living things which do not replicate, such as red blood cells, honey bees, and mules. Furthermore, there are non-living things which do replicate, such as crystals, fires, and asteroids.

It is not unusual for science to lack a definition for something. There are, for example, no scientific definitions for dark energy, the stock market, the causes of earthquakes, controlled thermonuclear fusion, the historical uniqueness of the ancient Greeks, or the fundamentals of human cognition and creativity – to name just a few of the unsolved mysteries of science. If we had a scientific definition for, say, the causes of earthquakes, we would then be able to successfully and repeatedly predict their occurrences just as we are now able to use our scientific definition for aerodynamics to accurately predict the characteristics of new aircraft designs – thank you Orville and Wilbur Wright.

In the absence of a scientific definition for life, biologists are left with the only alternative which is to provide a succinct, natural-language description.[76] This effort usually produces a fairly lengthy and complex result. Descriptions of contemporary living organisms normally include the following characteristics:[77]

- Homeostasis (internal control)

- Organization

- Metabolism

- Growth

- Adaptation

- Sensation

- Reproduction

While informative, such descriptions are nevertheless superficial. What's missing from this enumeration of features is the underlying commonality which ties all of them together. What singular set of molecular interactions accounts for both growth and sensation? If a scientific definition of life should one day be discovered, it would presumably provide an understanding of all of life's well known characteristics as derived side-effects from the elemental explanation.

A consequence of our ignorance as to the true nature of life is that it is impossible to describe the first living thing to appear on the earth. Clues are available, however, from what has already been discerned by science, including clues provided unwittingly by the deniers of evolution.

In taking a stand against the theory of evolution, the late astronomer Fred Hoyle once calculated the probability of assembling, by random acts of nature, the approximately two thousand enzymes necessary for all life on earth as one chance in ten to the forty thousand.[78] David Foster, a self-proclaimed "supernatural scientist" (whatever that means), calculated the odds of a virus, specifically the genome for the T4 bacteriophage, assembling spontaneously as one chance in ten to the seventy-eight thousand.[79] See Chapter 8 for an account of the spectacular size of a much, much smaller power of ten. These magnificent numbers are so gigantic they range beyond the abilities of the human imagination and are, for any practical purpose, effectively infinite.

While all life on earth today needs the two thousand enzymes mentioned by Fred Hoyle, it is not true that any particular cell needs all two thousand of them, nor is it known if the first living thing needed any of them. Also, biologists don't claim that the T4 bacteriophage was assembled all at once in a series of extraordinarily improbable random acts of nature, but rather it is assumed to be the result of billions of years of evolution wherein small random mutations occurred now and them.[80] But we will overlook these factual points since they tend to derail the train of thought required to deny evolution.

In the numerology of gigantic impossibilities, evolution-deniers have used these numbers, and others of a similar kind, to inform their audiences that nature could more easily assemble a fully functional passenger jet by spinning a tornado through a junk yard, and therefore, according to them, nature could not have created anything so complex as life.[81]

Oddly, evolution-deniers never seem to grasp the other side of these unimaginably huge numbers. Since it is so highly improbable that even the simplest known living cell could be assembled all at once in a spectacular series of random acts of nature, it is then highly probable that the first living thing to appear on this earth was far simpler than any such cell. The first airplane, after all, was far simpler than any present-day passenger jet. In fact it was so simple it could barely get off the ground, was dangerously unstable, flew only four times (all before noon on one day), covered a total lifetime distance of just a quarter of a mile, and was then wrecked by a mere gust of wind.[82] In the face of intellectually stifling statements that nature can't do it, let's amend the complexity of contemporary life with a few simplifying possibilities.

A better model for the first living thing is not the first airplane, but the first nuclear reactor, which was not made by man – it was made by nature, in Africa.[83] In the Oklo uranium deposit in Gabon two billion years ago, when the concentration of uranium-235 was higher than it is today, ground water seeping into the ore deposit slowed neutrons sufficiently to start a fission chain reaction. The subsequent release of more neutrons, as well as the creation of radioactive fission products, heated the water. When the water boiled away, the neutrons were no longer slowed down, halting the chain reaction. After the rocks cooled, water seeped back in and the reaction started up again. This cyclic process went on for hundreds of thousands of years with an average power output estimated to be around one hundred kilowatts. The very first living thing may have been a similar reaction – running under conditions which do not exist today, using available resources, then shutting down when those resources were exhausted, only to restart at a later time when better conditions returned.

There is no particular reason to believe that there ever was an identifiable "first living thing." If the fundamentals of life are in fact simple, then many different kinds of living things made up of different systems of molecules may have come into existence independently of each other. Many of these may have survived only a few hours, like the first airplane. Some may have destroyed each other on contact, some may have crossed paths without any interaction, and some may have developed symbiotic relationships. Communities of very simple "first living things" may have been created and destroyed billions of times in billions of places over hundreds of millions of years.

Just as the human body is a cooperative community of many different kinds of individual cells, none of them able to survive on their own but all of them providing for each other, so a single living cell may be seen as a cooperative community of many different kinds of individual molecular systems, none of them able to survive on their own but all of them providing for each other, and each of them a potential descendant from the long ago community of first living things. Analogs for all the structures and organs of multi-cellular plants and animals can be found at the molecular level in individual cells. While specialized cells in our bodies handle the functions of containment, movement, digestion, structure, defense, replication, repair, respiration, circulation, sensation, and decision making, so also do individual cells possess these capabilities via specific molecular systems. But there is no particular reason to believe that all of these specialized systems were actually necessary for life when life first appeared on the earth.

Many people, including many scientists, in fact including many biologists, remain inculcated with the idea that since almost all living things today are able to replicate themselves, self-replication must be a requirement for life. But the simpler the system the easier it is for replication to be accomplished by random acts of nature. For example, it rained. The rain caused water containing components of living molecular systems to spill over the banks of their pond and flow downstream into another pond. Or, having set up home in a crack in a rock, a surge of water can cause some of the assemblies of living molecules to flow into other cracks in the rock. In either case, this early form of life would have effectively been replicated into other locations by nothing more complex than a rain storm.

There is no particular reason to think that the first living thing needed to be protected inside a cell wall. It could have been thriving in a pond the size of an apartment building, or confined in a microscopic crack in a rock. Either way the local environment provides the container.

There is no reason to think that the first living thing needed any DNA to encode any information, or any RNA to transcribe it, or any molecular machinery of any kind to manufacture any other molecules. The molecules needed as resources for the life of the first living thing could have been provided by nature just as nature today provides us with water and minerals.

The evolution of the first, simple living thing into different "species" and/or into a greater complexity can also be accomplished by random acts of nature without requiring any DNA, RNA, or any manufacturing or replication processes. Other ponds, other cracks in the rock, can provide different environments with different concentrations of similar or different molecules. These new environments may even contain their own similar or different molecular living things. In any case, new environments create new opportunities which can alter the system of molecular reactions and produce either a more robust living thing, or a less capable living thing. Natural selection would weed out the more vulnerable performers while the more successful flourish.

All of this is, of course, conjecture, but it is the kind of conjecture which carries the hope of greater understanding. Until science discovers the definition for life, the evolution-denier's argument that life is just too complex to have been created by random acts of nature stands on a vacuous foundation. Absolutely nothing requires life to be complex, and as their own calculations indicate, life is far more likely to be simple. For all we know the re-creation of the first living thing may be occurring all over the earth today, but due to our ignorance as to the fundamental nature of life, we wouldn't recognize it if we saw it.

Finally, for those evolution-deniers who would rather not spend their time courting the rewards of scientific conjecture, there are dramatic biblical flaws in their argument which have already been mentioned but in the face of pernicious denial bare repeating. In particular, if life is too complex to be created by nature, then it is a miracle and therefore ubiquitous proof of God. But in this fallen world God has never provided proof of His existence with a miracle available all the time to everybody. He only creates miracles at His choosing for some people – the rest must nurture their belief with faith as the Bible clearly states in numerous places (see Chapter 18).

14

A Vignette from the History of Creationism

After years of trying to get their alternative to evolution legitimized, not through the presentation of any compelling scientific evidence, but by politicking public school boards with misinformation, the intelligent design movement received a crushing blow from the federal court on December 20, 2005.[85] In *Kitzmiller et al. v Dover Area School District et al.*, Judge John E. Jones III ruled that intelligent design is creationism – that it is religion and not science. As Judge Jones put it:

> . . . this case came to us as the result of the
> activism of an ill-informed faction on a school
> board, aided by a national public interest law
> firm eager to find a constitutional test case on

ID (Intelligent Design), who in combination drove the Board to adopt an imprudent and ultimately unconstitutional policy. The breathtaking inanity of the Board's decision is evident when considered against the factual backdrop which has now been fully revealed through this trial.[86]

Although it is unconstitutional to teach intelligent design as science in a public school, it can still be taught in private schools as well as in publicly funded courses covering other subjects, such as in a class on religion, or philosophy, or even politics. Regardless of fundamental legal issues arising from the foundation of the United States wherein a mandated separation divides state from church, a construct inspired by generations of European religious warfare, the state may also feel some obligation to protect its young citizenry from publicly funded fakery.

To movers and shakers throughout the widespread community of evolution-denial, the judgement from Pennsylvania could not go unanswered. The ruling struck at the heart of one's most significant beliefs, threatening to unhinged the life's work of numerous people fully committed to a path they are convinced is inspired and without fault. Nothing, therefore, could be more urgent than figuring out a way to seize the initiative, get back into the driver's seat, and roll on in defiance of all opposition. There were also immediate practical concerns regarding the judge's ruling. If they couldn't win their court case, they at least didn't want to lose any members of the choir.

After spending an entire year perusing the text of the ruling, evolution-deniers announced a profound discovery.

The judge, it seems, was a weak-minded fellow who had to resort to copying the work of other people.

> "Judge John Jones copied verbatim or virtually verbatim 90.9% of his 6,004-word section on whether intelligent design is science from the ACLU's proposed 'Findings of Fact and Conclusions of Law' submitted to him nearly a month before his ruling," said Dr. John West, Vice President for Public Policy and Legal Affairs at Discovery Institute's Center for Science and Culture.

> "Ironically, Judge Jones has been hailed as 'an outstanding thinker' for his 'masterful' ruling, and even honored by Time magazine as one of the world's 'most influential people' in the category of 'scientists and thinkers,'" said West. "But Jones' analysis of the scientific status of intelligent design contains virtually nothing written by Jones himself. This finding seriously undercuts the credibility of a central part of the ruling."[87]

Any lawyer exposed to only a handful of these words cannot be restrained from pointing out that Judge Jones was doing what every judge does. Both sides in a court case usually submit to the judge their separate proposals for the "Findings of Fact and Conclusions of Law." It simply saves time in our overworked legal system for the judge, if he agrees with the statements, to pick a draft of his ruling, or portions of his ruling, from the winning side.

The Discovery Institute acknowledges as much, yet they nevertheless insist that the numerical information they have wrung out of six thousand (and four) words from a judicial ruling totaling more than thirty four thousand words represents sufficient evidence of incompetence, or sloth, or both. The obedient members of the choir can now absorb the score of this new tune – that Judge Jones and his infuriating decision are irrelevant to any serious discussion on the veracity of intelligent design. All propagandists begin their day with the firm belief that the average person has neither the intelligence nor the experience nor the desire to perceive the truth, and even if they do, they have neither the wisdom nor the will to act on it.

But bigger ideas than spiteful press releases were in the works. Personal experiences of great events from long ago can be remembered with clarity as though they transpired only yesterday, while many other events which did in fact occur yesterday can barely be recalled. For evolution-deniers, the judgement from Dover needed to be set into the mind's faded memories by engineering the sunrise of a new great event.

In March of 2003, at the 75th Academy Awards ceremony, the Academy of Motion Picture Arts and Sciences announced that the winner for Best Documentary of 2002 was "Bowling for Columbine," directed by Michael Moore. A year later at the Cannes Film Festival, Mr. Moore received the most prestigious European film award, the Palme d'Or, for his documentary "Fahrenheit 9/11." Hollywood's Oscar-winning redefinition of the word "documentary," seconded by the artistic elite of Europe in a twenty minute standing ovation, provided an expanded meaning to factual accounts, now including the skilled misinformation of any crowd pleasing, image-speak propaganda.[88] With the standards for truth lowered, the designers of the anti-evolution edifice enthusiastically embraced all the possibilities of this new fashion. A world premiere revitalization of their cause needed only money, guile and hard work – all of which they had in abundance.

Predictably, the first step began with deception. A production company called Rampart Films claimed to be making a documentary to be titled *Crossroads: The Intersection of Science and Religion*. The purpose of this film was to present an objective and balanced treatment of issues on the two subjects, or so it was assumed.[89] But in fact these elements were the principle components of a façade which was used by representatives of the project to initiate interviews with many well known defenders of the theory of evolution, namely:

- Peter Atkins, chemist

- Richard Dawkins, biologist

- Daniel C. Dennett, philospher

- P. Z. Myers, biologist

- William Provine, science historian

- Michael Ruse, philosopher of biology

- Eugenie C. Scott, Director of the National Center for Science Education

- Michael Shermer, Executive Director of the Skeptics Society

Not only are all of these individuals well known for their defense of the theory of evolution, they also happen to be well known as avowed atheists. Noteworthy, therefore, is the fact that in the pantheon of public evolution defenders no one who believes in both God and evolution, such as biologist Kenneth Miller, was interviewed for the film. When asked about this apparent oversight, associate producer Mark Mathis said it was no oversight. The inclusion of Miller, he said, "would have confused the film unnecessarily."[90]

Excluding Miller does indeed help clarify one of the false messages of the film – that those who defend the theory of evolution are godless people and that the acknowledgment of evolution as the origin of species inevitably leads to atheism. The facts left out of the film speak otherwise. The Clergy Letter Project, which accumulates a list of supporters of the theory of evolution from clergy representing numerous denominations, has many thousands of signatures.[91] With a myopic vision few have ever achieved, it remains lost on the makers of this film that expelling Miller also clarifies the message that the works of Christian evolution-deniers are designed from the get-go to be manipulative and deceitful.

Having acquired the video clips, sound bites and release forms to cut and paste their way around the center ring of their new, big tent, the false-pretensers excused themselves of all their false pretenses, dropped the façade, and with money that could move mountains, Premise Media, the real production company, launched into more than a thousand theaters *Expelled: No Intelligence Allowed*.[92] This was to be their new place in the sun, the clearest and brightest statement they could ever make on their legitimacy as a cause.

Beginning with a lengthy journey through the diminished careers of academics famous for their unprofessionalism, such as the case of Richard Sternberg,[93] or denied advancement for their lack of performance, such as the case of Guillermo Gonzalez,[94] the audience has to wait almost an hour before the judgement from the Dover courtroom is addressed. In his on-screen interview William Dembsky finally lays out the new mantra:

I think court cases don't decide anything.[95]

Trying to dodge a bullet after it hits doesn't count for much, but here at least we're reminded that we are always free to sweep our troubles away with denial.

Aside from misleading commentary on careers, atheism, and the power of the federal court, the film also makes the blunt contention that Charles Darwin and his theory of evolution are culpable for the massive crimes of Adolf Hitler and the Nazi Third Reich. This revelation is introduced by David Berlinski, whose academic achievements include a Ph.D. in Philosophy from Princeton University, followed by postdoctoral research in both mathematics and molecular biology at Columbia University. Dr. Berlinski has written numerous books on mathematics, as well as a series of detective novels, and has held teaching positions in English, mathematics, and philosophy at many universities, including, but not necessarily limited to, the City University of New York, Rutgers, San Francisco State University, San Jose State University, Stanford University, the Université de Paris, the University of Puget Sound, the University of San Francisco, the University of Santa Clara, and the University of Washington. According to the capable and accomplished

Professor Berlinski:

> . . . if you open *Mein Kampf* and read it, espe-
> cially if you can read it in German, the corre-
> spondence between Darwinian ideas and
> Nazi ideas just leaps from the page . . .
> Darwinism is not a sufficient condition for a
> phenomenon like Naziism, but I think it's
> certainly a necessary one.[96]

It should come as no surprise to any ordinary person that Professor Berlinski has never been asked by any university to instruct a class in history. To imagine that Darwin's ideas were "necessary" for Naziism, that without the theory of evolution Hitler would never have been able to launch a continental, racist massacre of millions of helpless people, defies both common knowledge and everyday rational thought. Such a dunderheaded statement demonstrates both ignorance of history and ignorance of human nature. On the other hand, perhaps Dr. Berlinski's commentary is just another example of the everyday thoughts from a cynical, misinformation voyeur.

Acts of mass murder launched by the Nazis trace neither their source nor their motivation to Charles Darwin and *The Origin of Species*. Such a naive, or manipulative, contention does not acknowledge the truth – it disguises it. The roots of the Nazi attack on the Jews are easily revealed as both ancient and well known, seen especially in the very long European history of violent Christian anti-Semitism. Here, from over the last nine hundred years, are a few examples underpinning and foreshadowing the Nazi horror:

- In 1094, as a leader in the First Crusade, Godfrey of Bouillon proclaimed that he would "avenge the blood of Jesus on that of the Jews, and leave none of them alive." But his companions were more generous, declaring that they would exterminate the Jews, unless they converted.[97]

- In 1190 a wave of violence against Jews swept across England, including the slaughter of the entire Jewish community of York at Clifford's Tower on March 16.[98]

- In November of 1215, at the Fourth Lateran Council, which assembled from across Europe dozens of patriarchs, hundreds of bishops, plus envoys from Emperors and Kings, Pope Innocent III issued seventy decrees, or Canons, governing the lives of Europeans in the Middle Ages, including a decree forcing Jews to wear a yellow badge so they could be distinguished from Christians.[99]

- On July 18, 1290, Edward I (Longshanks) ordered the expulsion of the Jews from England, seizing for himself the property they could not carry.[100]

- In 1306 Philip the Fair expelled the Jews from France, seizing all their properties for himself, including what they could have carried.[101]

- In the spring of 1348 the Black Death began its annihilating march across the European continent, stopping whole armies in their tracks and providing a new pretext for scapegoating Jews. Overlooking the plain fact that Jews suffered from the plague just as much as their Christian neighbors, a point made by Pope Clement VI as well as other rulers in vain attempts to stem the violence, accusations nevertheless flourished among the common people that the huge death toll was caused by Jews poisoning the water supplies in an organized bid to destroy all of Christendom. Proof of this conspiracy was successfully extracted here and there by torture. In more than three hundred cities and towns across Switzerland, Germany and Austria, massacres of Jews preceded the onslaught of the Death. On January 9, 1349, in Basel, Switzerland, six hundred shackled Jewish men and women were burned alive in a wooden barn constructed by the gilds specifically for that purpose. A few weeks later, in Strasbourg on February 14, two thousand Jews were dragged to the burial ground and burned en masse. Four months later the Scourge hit Basel and Strasbourg like a couple of neutron bombs, sparing the buildings while killing thirty thousand people.[102]

- In 1543 Martin Luther issued his infamous treatise, "On Jews and their Lies" where one can find advice on setting fire to synagogues and burying whatever will not burn, razing Jewish homes, the forced labor for "young, strong Jews and Jewesses," a brief note about being "at fault in not slaying them," and this typical rant:

"I have read and heard many stories about the Jews which agree with this judgment of Christ, namely, how they have poisoned wells, made assassinations, kidnaped children, as related before . . . For their kidnaping of children they have often been burned at the stake or banished (as we already heard). I am well aware that they deny all of this. However, it all coincides with the judgment of Christ which declares that they are venomous, bitter, vindictive, tricky serpents, assassins, and children of the devil who sting and work harm stealthily wherever they cannot do it openly . . . That is what I had in mind when I said earlier that, next to the devil, a Christian has no more bitter and galling foe than a Jew. There is no other to whom we accord as many benefactions and from whom we suffer as much as we do from these base children of the devil, this brood of vipers."[103]

- On May 26, 1753, the Bishop of Kiev helped secure the public execution of fifteen or so Jews in Zhitomir, Russia, for a crime they didn't commit, including that one couple who converted to Christianity on the spot and were thereby granted the grace of a beheading instead of being quartered alive like the others.[104]

The list goes on and on – but rather than consume hundreds of pages with documented repetitions of this overwhelming cruelty, let's just fast forward through countless years of mass evictions, rapes, tortures, murders, barbaric public executions, and business inspired, government organized massacres of countless, helpless men, women and children, and brake to a halt at the beginning of Hitler's Third Reich, where this long legacy of Christian anti-Semitism provides far more than just an architectural sketch for industrialized mass murder.

In speaking of a correspondence between ideas that "just leaps from the page," Dr. Berlinski somehow neglected to mention the closing remarks from Chapter 2 of *Mein Kampf*, "Years of Study and Suffering in Vienna," wherein Hitler assails us with:

> If, with the help of his Marxist creed, the Jew is victorious over the other peoples of the world, his crown will be the funeral wreath of humanity and this planet will, as it did thousands of years ago, move through the ether devoid of men.

> Eternal Nature inexorably avenges the in-
> fringement of her commands.
>
> Hence today I believe that I am acting in
> accordance with the will of the Almighty
> Creator: by defending myself against the Jew,
> I am fighting for the work of the Lord.[105]

Can we assign blame to Charles Darwin for the inspiration behind, "Eternal Nature inexorably avenges the infringement of her commands?" Why not Isaac Newton? But then to whom do we assign blame for "in accordance with the will of the Almighty Creator?"

Like countless other dictators throughout history, Hitler was always more than ready to vent his spleen into a boiling verbal brew of whatever terminology got him the results he wanted – Marxist, Jew, Eternal Nature, Almighty Creator, Lord. But when all those Nazis true-believers dragged Europe through Hell, the belt buckles of the Hitler Youth, the Police, and the entire Wehrmacht (army, navy, and air force) were not inscribed with "Darwin mit uns." They were inscribed with "Gott mit uns" (God is with us).

By turning a blind eye to the deep religious, political, economic and cultural roots of the Holocaust, this film exposes itself as a defiant, self-serving distortion of reality toying with the same contempt for the truth that made Nazi propagandists the cheerleaders of global infamy. While some people grab for attention by denying the Holocaust, here we have a group of people attempting to hijack it for their own purpose. Using one of history's most terrifying atrocities for the bizarre expediency of passing off its crimes onto a recent theory of science is an insult to every victim of Nazi brutality, especially in the face of a relentless Christian fanaticism which laid the foundations of the Holocaust across the long centuries before Charles Darwin was even born.

But Dr. Berlinski is not the only one in this film trying to prop up hideous Nazi crimes with straws from Darwin. Taking on a new career with the pretensions of an honest intellectual investigator and fact finding adventurer, as well as one of the script writers for this movie, Ben Stein leads the audience around the ring of half-truths like a bull by the nose. For example, while supposedly exposing the source of Nazi eugenics, Mr. Stein reads to us from Darwin:

> With savages, the weak in body or mind are soon eliminated. We civilized men, on the other hand, do our utmost to check the process of elimination; we build asylums for the imbecile, the maimed and the sick, thus the weak members of civilized societies propagate their kind. No one who has attended to the breeding of domestic animals will doubt that this must be highly injurious to the race of man. Hardly anyone is so ignorant as to allow his worst animals to breed.

> Charles Darwin, *The Descent of Man*, eighteen seventy one.[106]

This somber reading is then followed by a full ninety seconds devoid of any dialogue, perhaps to allow plenty of time for the dark message to really sink in. Apparently we are to believe that Charles Darwin favored the breeding of a superior race of man by culling the human population of its weaker members before they can produce more of their imperfect kind. What is astonishing about these words is not that Charles Darwin wrote them, which he certainly did, but that the producers of this film would be so arrogant as to think – in the age of the Internet – that people could not easily check for themselves whether or not Ben Stein sought to deceive his audience by deliberately rigging this quote. Here is what Charles Darwin actually wrote, with all the words from Ben Stein's reading underlined, and all the rest expelled from his script:

> <u>With savages, the weak in body or mind are soon eliminated</u>; and those that survive commonly exhibit a vigorous state of health. <u>We civilised men, on the other hand, do our utmost to check the process of elimination; we build asylums for the imbecile, the maimed, and the sick</u>; we institute poor-laws; and our medical men exert their utmost skill to save the life of every one to the last moment. There is reason to believe that vaccination has preserved thousands, who from a weak constitution would formerly have succumbed to small-pox. <u>Thus the weak members of civilised societies propagate their kind. No one who has attended to the breeding of domestic animals will doubt that this must be highly injurious to the race of man.</u> It is surprising how soon a want of care, or care wrongly directed, leads to the degeneration of a domestic race; but excepting in the case of man himself, <u>hardly any one is so ignorant</u>

<u>as to allow his worst animals to breed.</u>

The aid which we feel impelled to give to the helpless is mainly an incidental result of the instinct of sympathy, which was originally acquired as part of the social instincts, but subsequently rendered, in the manner previously indicated, more tender and more widely diffused. Nor could we check our sympathy, if so urged by hard reason, without deterioration in the noblest part of our nature. The surgeon may harden himself whilst performing an operation, for he knows that he is acting for the good of his patient; but if we were intentionally to neglect the weak and helpless, it could only be for a contingent benefit, with a certain and great present evil.[107]

With all its untrustworthy dialogue, in-your-face imagery and manipulative background score, the film hammers home the tacit message that if an idea, such as evolution, can be linked to horrendous crimes, then that idea should also be condemned for those crimes. If this is an example of wisdom, where else does this light fall? Does it fall on more than seventeen hundred years of a ferocious persecution of Jews by Eastern Orthodox, Catholic, and Protestant Christians, all of whom were more than ready to state their justifications? Does it fall on James Warren Jones (a.k.a, Jim Jones), the founder and leader of the "Peoples Temple Christian Church Full Gospel" and the preacher and practitioner of "revolutionary suicide?" If so, then we can reach past this particular maniac to hold Christianity culpable in the deaths of about nine hundred men, women and children.[108] But to avoid a long history littered with lunatics, fiends, and manipulators who have used Christianity to accomplish their nefarious goals, simply adopt the evolution-denier's tunnel vision and stare fixedly at Charles Darwin – perhaps no one will notice that the behavior is two-faced.

While evolution-deniers may have fashioned their film from contemporary inspirations, their creation also exhibits a well known style famously articulated in 1925:

> In this they proceeded on the sound principle that the magnitude of a lie always contains a certain factor of credibility, since the great masses of the people in the very bottom of their hearts tend to be corrupted rather than consciously and purposely evil, and that, therefore, in view of the primitive simplicity of their minds they more easily fall a victim to a big lie than to a little one, since they themselves lie in little things, but would be ashamed of lies that were too big. Such a falsehood will never enter their heads and they will not be able to believe in the possibility of such monstrous effrontery and infamous misrepresentation in others; yes, even when enlightened on the subject, they will long doubt and waver, and continue to accept at least one of these causes as true. Therefore, something of even the most insolent lie will always remain and stick . . .
>
> *Adolf Hitler*[109]

The real problem with lies, as most of us have long since found out, is in their great potential for self-injury. Like boomerangs, they are thrown to accomplish some goal, but they can come back at you. *Expelled: No Intelligence Allowed* takes self-injury to a much deeper level. Instead of a boomerang, the misrepresentations and lies are more on the scale of a wrecking ball. They have swung it out to knock down the edifice of evolution, but it comes back at them. Documentaries, as with any art form, don't just speak of their subjects – they also speak of their creators. This film was created by a group of people so clueless of their own hypocrisy that they end up lauding themselves in public with breathtaking inanity.

15

The One-Paragraph Proof of Evolution

In the creationist film *Expelled: No Intelligence Allowed*, Dr. David Berlinski claims that the theory of evolution is nothing more than smoke:

> One of my prevailing doctrines about Darwinian theory is, man, that thing is just a mess. It's like looking into a room full of smoke. Nothing in the theory is precisely, clearly, carefully defined as delineated.[110]

For generations evolution-deniers have portrayed the gaps in the fossil record as a decisive flaw in the theory of evolution. Some claim that the gaps are clear evidence for the *ex nihilo* (out of nothing) creation of organisms by God, or by "the Designer," as they say at intelligent design events. Others claim the gaps expose evolutionary scientists as aggressive atheists promoting an anti-religious theory which could not possibly be supported by the fossil record's supposedly paltry assortment of highly fragmented samples. If the ultimate goal of a Christian education is an uneducated congregation, then these smokescreens have proven their worth in hiding the truth about the gaps in the fossil record.

Science can't speak to divine creation since there is no physical evidence of it, but it can speak to the evidence these gaps present and that evidence makes one of the most powerful statements in history. The gaps in the fossil record provide the simplest and most spectacular validation of the theory of evolution. Here, in one paragraph, is the proof of evolution, followed by four classic examples delivering that proof.

All scientific theories provide their most dramatic confirmation by their ability to predict the future – an ability they are required to demonstrate time and time again.[111] You wouldn't normally think of the theory of evolution, so thoroughly rooted in its dusty past, as the proverbial crystal ball, but it is. Only the theory of evolution lays out the fossil record, which it presents as a map of the history of life. Since this record is hugely imperfect, for all the biological and geological reasons Darwin so carefully explained, there are inevitable gaps.[112] If evolution does indeed render an accurate description of the history of life, then these gaps can be used to foretell the future. They tell paleontologists what kinds of fossils they will one day discover. Sometimes the gaps can even provide clues as to where on the earth the predicted fossils can be found. Here are examples of the theory of evolution, successfully and repeatedly, predicting the future:

- Fossils of the sensational *Archaeopteryx*, a dinosaur-bird transitional animal, were first discovered in Germany in the early 1860's, shortly after the publication of *The Origin of Species*.[113] This animal sits in the gap between small theropod dinosaurs and modern birds – much closer to dinosaurs than to birds. It is defined as transitional because it has many features found in dinosaurs which are not found in birds, such as teeth, as well as a few features found in birds which are not found in dinosaurs, such as an opposable big toe. The fact that it also exhibits wings and feathers further characterizes its transitional nature. That this fossil even exists at all leads to several evolutionary predictions. First, other similar transitional fossils of small, dinosaur-like birds and bird-like dinosaurs are likely to be found one day. Second, when these predicted fossils are discovered, they will appear in a geologic time frame clearly associated with *Archaeopteryx*. And finally, as more and more of these fossils are inevitably found, they will increase the resolution of our understanding of the evolutionary development of modern birds. All of this is precisely what has happened with the many recent discoveries from fossil beds in northeastern China.[114]

- In 1962 E. J. Slijper noted that not a single transitional fossil existed between the earliest known example of a whale, from about 45 million years ago, and their presumed ancestors – four-legged, hoofed land animals from millions of years before that.[115] Since 1962, however, field work in Pakistan and Egypt has brought to light a series of fabulous transitional fossils which have fulfilled the predictions of evolutionary scientists.[116] One of the most significant, from around 49 million years ago, is *Ambulocetans natans*, a whale-like animal adapted to both fresh and salt water. Like whales and those land-based ancestors, its ears were not external – it could hear sounds by picking up vibrations through its jawbone. Though clearly a swimmer, it also had legs, each toe of which terminated with a miniature hoof.

- Recently a tiny bee was found in a piece of hundred million year old amber. You can find an article about it from October 25, 2006 on the BBC News website.[117] The entombed bee had some features found in wasps, fulfilling the prediction that bees evolved from wasps around the time of the great angiosperm radiation (the spread of flowering plants) which occurred during the Cretaceous, approximately 100 million years ago.[118]

- Ancient rock composed of freshwater sediments 380 million years old contains the fossils of lobe-finned fish, but no fossils of amphibians. The first amphibians are found in freshwater sediments beginning 363 million years ago. From this information it's effortless to predict that a fossil of a transitional animal on the evolutionary pathway from fish to amphibians, from fins to limbs, should be found in rock made from freshwater sediments that is between 380 and 363 million years old. A group of scientists took a geological map of North America and identified a barely accessible area in the Canadian Arctic where the appropriate rock type of the appropriate age was exposed. After several years of digging they discovered, in 2004, exactly what they expected to discover – *Tiktaalik roseae*, an animal that lived 375 million years ago with characteristics of both fish and amphibians.[119]

Many more examples are available to anyone pursuing an education in the subject, including the most spectacular example of all – the successfully predicted evidence for the evolution of man.

16

Freedom of Thought

Backed into a corner by evidence and on fire with denial, claiming that the fossils are just a pile of bones with no history to tell, it is not uncommon for evolution-deniers to end discussion with the walk-out statement, "You can't prove it!" As a matter of fact no one can really prove anything to anybody since we are always free to pick and choose from among our own thoughts. But this freedom of thought is no safe harbor for evolution-deniers who also happen to be Christians.

Of all the two-edged swords around, freewill is the most famous, and the most infamous. The ease with which this blade cuts through the boundary between good and evil is legendary. While enabling some demonic overachievers to cause a world of pain, it also provides each of us with the means to save ourselves or ruin ourselves, as the case may be. Yes, anyone can freely choose to believe whatever they want. For a Christian however, the Bible is all about defining what the right choices are.

As a Christian there are limits to what you can choose to

believe without tripping over the stumbling blocks of iniquity, or worse – becoming one yourself. For example, if you choose to reject the predictive proof of the theory of evolution, then you will have to also reject all other theories in science which are proven the same way, otherwise you would not be fair. This means, to be fair, you would have to reject all of science.

You will, for instance, have to reject the existence of magnetism and x-rays, both of which we will never see since we are physically incapable of seeing them – we only demonstrate their existence by their effect on things which we can to see, and "to demonstrate" is to fulfill a prediction.

17

The Debate

The argument for intelligent design is simple and straightforward. Just as the amazing complexity of a computer implies the handiwork of a skilled electronic engineer, so then does the even greater complexity of a living organism imply the talents of an extraordinarily capable biological designer. An attempt is made to strengthen this argument by appealing to your sense of what is plainly obvious. Living organisms are known to be made up of very intricate molecular machine-like components using geometries and forces no human being can command and it seems, therefore, more than obvious that these mechanisms must have been fashioned by a craftsman with superhuman abilities. Numerous negative statements about the theory of evolution are also provided in the hope of strengthening this argument even further.

Overlooking the fact that many of the negative statements about evolution are flat-out lies and easily shown to be so, in principle, negative statements about the theory of evolution do not support intelligent design in the same way that negative statements about the President of the United

States do not support any opponents. The opponents, after all, could be worse.

The appeal to obviousness is not impressive to scientists for the simple reason that anyone can find examples in which the truth is not obvious. Obviously it was a great looking apple until after the first bite and half a worm was found wiggling in what was left. From history we have striking cases of whole civilizations, one after the other, being fooled by the obvious. It is plainly obvious, for example, that the entire universe revolves around the earth, but we now know that this is not true.

The argument for intelligent design is an analogy which makes the claim that there are profound similarities between the machines made by man and the fabulous molecular mechanisms operating in living organisms. From this the conclusion is offered that living organisms must also be intelligently made. Scientists, however, don't find this analogy very compelling.

Analogies are one of the easiest ways to misinform people and thus provide one of the worst foundations for a scientific explanation. Even Darwin had something to say about it – "But analogy may be a deceitful guide."[120] For example, just as birds flap their wings to ascend, so then should the first flying machine be made to flap its wings in order to do likewise.

The promoters of intelligent design also make the baseless claim that a machine can only be made by an intelligent being. There are in fact many machines in nature, some of them so complex we still don't fully understand how they work. Here is a list of a few all-natural, non-biological machines:

- The Gulf Stream and the Mississippi river are naturally occurring conveyor belts which we use to facilitate our commerce.

- The entire solar system is an elaborate clock – in 1676 Jupiter and its inner moon Io provided the timepiece used in the first measurement of the speed of light.[121]

- The sun is a light bulb.

- The weather on occasion replaces the function of a lawn sprinkler.

- A crystal is a quantum-mechanical device which relentlessly manufactures copies of itself – a functionality essential to many of our industries.

- Two billion years ago the Oklo uranium deposit was a nuclear reactor which operated for thousands of centuries (see Chapter 13).

- Antarctica is a refrigerator complete with insulation in the form of the Antarctic Circumpolar Current which protects the ice sheet from waters warmed in the tropics.[122]

- A galaxy cluster is a relativistic telescope with a gravitational lens that can bring into view objects in the far distant universe.

Every one of these mechanisms is an ordinary product of the

laws of nature, as are the complex molecular machines found in living cells (see Chapter 6).

Such counter statements, however sensible they may be, do not impress the dedicated deniers of evolution. In fact they love this stuff – it's exactly what they want to hear. Debates between scientists and evolution-deniers serve to create an illusion of equality where in fact there is no such equality. Like a magician drawing your attention away from the sleight of hand, it's far better for the apologists of anti-evolution to run on and on over the fine points of the above criticisms than to risk revealing the real reason intelligent design is an unmitigated scientific failure.

The theory of evolution has made innumerable predictions which have come true, just like many other hugely successful scientific theories. Furthermore, there are currently countless predictions in the docket waiting fulfillment. In contrast to this, the so-called "theory of intelligent design" has predicted in the past, and currently predicts today, absolutely nothing. All the made-up controversies regarding evolution stand meaningless in the light of this single catastrophic failure.

It appears that God, who is infinite and therefore possesses infinite intelligence as well as infinite ability, presumably including talents our puny minds are thoroughly incapable of comprehending, has established a way to remain unseen in this fallen world, except of course through His measured revelations, by, among many other things, creating the ultimate achievement in creativity – a creation which is itself creative.

18

Pick a Universe

At the heart of the Christian denial of evolution is the refusal to accept the evolution of man within the order of primates (from the Latin *primas*: of first rank). The idea that when man first thought of God, God had already thought of man, is effortless to comprehend. So it is of no surprise to anyone that at the very beginning of man's relationship with God, as revealed in the Bible which has been compiled from the beginning of history and dedicated thoroughly to the connection between man and God, the creation of man is described by the inspired text in the terms of this sacred relationship.

> Then the Lord God formed man of dust from
> the ground, and breathed into his nostrils the
> breath of life; and man became a living being.
> *Genesis 2:7*

Science catalogues the evidence that we were created by nature from the stock of primates, but the Bible says we were fashioned from dust by the hands of God. Is the correct description of our origin either one or the other, or are both accounts correct?

While computers are restricted by their logic engines to a "world view" consisting of only true or false, one or zero, the human mind, which according to the Bible was formed in the image of God, can range beyond this unconditional simplicity. Our capacity for creative thought is grounded in an ability to recognize objects, events, and conditions which are simultaneously true and false. For example, Shakespeare's famous description of his female protagonist in *Romeo And Juliet* thankfully does not include any of the clinical accuracy of science, yet as almost everyone older than twelve knows, it is perhaps impossible to provide a more accurate depiction of a young woman, especially as seen through the eyes of a young man:

> But, soft! what light through yonder window
> breaks? It is the east, and Juliet is the sun![123]

From a purely scientific point of view, Romeo's answer to his own question is ridiculously false, and yet it is so perfectly the truth.

To think that God is not also capable of making creative statements which are simultaneously true and false is of course an absurdity. One only has to consider the many parables told by Jesus to recognize that the Genesis portrayal of God's creation of man can be acknowledged as true without denying any of the factual precision of its scientific explanation.

There are of course numerous Christians who believe in a literal interpretation of the Bible. While the strictness of this view is usually mentioned only in reference to the Old Testament, obviously the New Testament is included in their interpretation. They can therefore be counted on to be foremost in reminding their fellow Christians that by accepting Jesus everything is made new, and that they are free from the rote stricture of the old Law, and that no justification exists for deceit about science – not even the thought of it.

Any wisdom derived from these considerations is rejected by evolution-deniers. Like a failing firework, their inventions spew out a litany of flashy debris – science leads to atheism, science is just another philosophy of mere personal opinions while evolution is "just a theory" of mere speculations, scientists are contemptuous of religious beliefs and operate an academic conspiracy to destroy the careers of religious colleagues, evolutionary scientists in particular are running a scam just to hang onto their academic salaries, science is a fundamentally evil force in society and needs to be corrected by recognizing the authority of God (whatever that means), and so on and so on. In all of this, however, the consequences of throwing away the theory of evolution and replacing it with intelligent design are never presented. What would our world actually be like if the leaders and followers of anti-evolution got what they wanted? Imagine, for example, two parallel universes – the universe as portrayed by science, and the universe of intelligent design.

Science does not say that God does not exist, as some mis-informants claim. Instead, science deliberately views reality with God removed, and for good reason. By ignoring the supernatural, which is after all not accessible to any reliable investigation, and by focusing instead on only the repeatable characteristics of nature, we have discovered that an understanding of the mechanisms of nature can lead to dependable and accurate predictions. Along with the sublime reward of unveiling the nature of nature, these scientific achievements have given us access to a broad range of extraordinary capabilities.

In its thoroughly all-natural view, the universe as described by science is a vast structure with no apparent purpose. By definition it is the random creation of some known, some suspected, and certainly some still unknown, physical factors.

Interestingly, by a loosened definition of the length of the first days in the Book of Genesis, an alignment can be seen between the Genesis account and the Big Bang theory of the origin of the cosmos. The Creator's statement, "Let there be light," could hardly be made more fittingly than with the Big Bang. This ostensible alignment is often mentioned by some groups of intelligent design advocates as scientific evidence for the accuracy of the Bible. The claim is clearly based on viewing Genesis as an obscure, rather than a literal, description of origin since a few "days" of creation must be stretched across billions of years of cosmological history. Yet these same thinkers then switch to being literal when man is created, indicating that their interpretation is in fact a patchwork stitched together with the threads of inconsistency. Given, however, the historically capricious nature of cosmological theories, the obscure alignment is very likely a fluke and very unlikely to last.

Since the cosmos is, by definition, infinite, all options for its origin are on the table at all times. The history of cosmological theories is a history of revolutions. Whatever the accumulated evidence might point at today, additional evidence may not point there tomorrow. Any indications of any event occurring before the Big Bang, or any structure existing beyond what we currently perceive as our "universe," or any new insights into the underlying physics of the cosmos, can radically alter the current concept with nature's creative violence then being conducted in a much larger, or at least much different, volume.

In this immense universe of no discernable purpose, science classifies us also as purely random creations of nature. Not only has diverse and repeating evidence clearly established that we evolved from more primitive primates, but more than that, according to the theory of evolution we have common ancestors relating us to every single living thing, past and present, including the bottom-dwelling, slimy sea slug, the ancient and beautiful ginkgo tree, the majestic blue whale, and every individual, inconsequential blade of grass. We are even distantly related to *Yersinia pestis*, the bacteria which causes the bubonic plague. It appears that life on earth has organized itself, thanks to the steady supply of energy from the sun, into a gigantic and highly interacting ensemble of DNA-controlled constructions of which we are only one of the most recent and most highly refined. And as if these simple facts were not soulless enough, in this mechanistic universe of science, constantly jarred by nature's random events across a vast range of magnitudes, not only is there no scientific evidence for the existence of God, there is every reason to expect that there will never be any such evidence.

In contrast, the universe of intelligent design is filled with many varied works conspicuously made just for us by the Father. This comforting view is not your great-great-great-grandfather's faith-based intelligent design since this modern rendition takes the extraordinary step of claiming scientific

124

authentication. As Philip E. Johnson put it:

> My colleagues and I speak of "theistic real-
> ism" -- or sometimes, "mere creation" --as the
> defining concept of our movement. This
> means that we affirm that God is objectively
> real as Creator, and that the reality of God is
> tangibly recorded in evidence accessible to
> science, particularly in biology.[124]

In the Designer's universe, Charles Darwin and his theory of evolution reside in the dustbin of history. Darwin's theory is seen as a wrong-thinking scientific obsession which damaged society for around a hundred and fifty years, but was finally replaced thanks to the brilliant work of more enlightened people. Michael Behe wins the Nobel Prize for his discovery of the fatal flaw in evolutionary theory, namely that the intricately fashioned molecules making up the internals of living organisms are just too complex to have been created by nature. William Dembsky, in a virtual repeat of Einstein's long but ultimately successful struggle with gravity, finally produces a positive definition of design to replace his negative, and useless, "Explanatory Filter," and for this achievement he receives his own Nobel.

As a result almost all scientists accept the theory of intelligent design. There are, of course, a few irrelevant contrarians around, but they're easily dismissed as just another example of the everyday failings of human nature. The world's biologists are fully aware that intelligent design is now the only scientific theory which makes no predictions whatsoever, but that's entirely acceptable because the very nature of God's activities are unpredictable.

Intelligent design is taught in high school science classes and, due to the easily provable impossibility of evolution, it is recognized as the only valid explanation for the history of life. There are still other religions in the world claiming that their god, or gods, did it, so that conflict continues. In response to this, the heroes of intelligent design finally stop the obvious sin of disguising God with the term "Intelligent Designer", something the twelve Apostles would never have done while living and dying for the Word, and they are, of course, forgiven by their peers for this evidently necessary subterfuge.

With the scientific acceptance of intelligent design, the number of Christians in the world skyrockets. Almost all scientists become Christians. Christian pastors now have a fantastic resource for dealing with doubt and unbelief. In fact, doubt about God effectively disappears since it can now be shown scientifically that He exists. Anyone who has any doubt or unbelief can simply take a course in biochemistry of all things and with any average microscope they can see the truth for themselves in the frenetic tentacles extruded by bacteria. Those who like a more cerebral challenge can instead study Dembski's rarefied mathematics – famous for its shape-shifting terminology and an almost complete suffocation of clarity.

In the universe of intelligent design, God has placed His unique signature in the complexity of the molecules which lie at the foundation of life. All living things carry this signature. With no natural explanation for the design of these life-giving molecules, their impossibility infallibly proves the existence of God. Moses had his burning bush, but that was only for him. The Jews escaping enslavement in Egypt experienced an awesome menagerie of signs and wonders – the

Red Sea miracle, the pillar of fire, manna from heaven, and all the others, but those were only for them. The very first Christians were blessed with the greatest miracle of all – the defeat of death by the risen Christ, but that startling reality was for their eyes only. Until the theory of intelligent design was accepted by the scientific community, believers had to have faith that God exists, but now, for the first time in history, they actually have direct, physical proof. It is finally acknowledged that God has placed an indisputable burning-bush-for-everyone in every single living thing.

It is clear what the proponents of intelligent design and the deniers of evolution and all their good Christian follow-ers want. They talk about it whenever the subject of science comes up. They just can't wait for that day when the whole world finally accepts the fact that the theory of evolution belongs in the trash. But this stand has always been taken to fulfill a vision of what *they* want – a vision of a universe and a reality which *they* desire. Conveniently ignored in all of this, however, is any real consideration for what God might want.

> "For My thoughts are not your thoughts, neither are your ways My ways," declares the Lord.
>
> *Isaiah 55:8*

Which of these two universes – the universe of no scientifically discernable purpose, or the universe of intelli-gent design, best fits *His* goals? Perhaps the Bible contains some clues.

To say that the Bible reveals God's desire for His people to have faith is more than a slight understatement. Here are a few examples:

> And without faith it is impossible to please Him, . . .
>
> *Hebrews 11:6*

> For by grace you have been saved through faith; and that not of yourselves, it is the gift of God;
>
> *Ephesians 2:8*

> for we walk by faith, not by sight –
>
> *Second Corinthians 5:7*

> Therefore having been justified by faith, we have peace with God through our Lord Jesus Christ, through whom also we have obtained our introduction by faith into this grace in which we stand; and we exult in hope of the glory of God.
>
> *Romans 5:1-2*

> Behold, as for the proud one, his soul is not right within him; but the righteous will live by his faith.
>
> *Habakkuk 2:4*

Although most people would admit to being somewhat unfamiliar with the Book of Habakkuk, it is of no minor significance that the last line above is repeated in the New Testament three times:

Now that no one is justified by the Law before God is evident; for, "the righteous man
shall live by faith."

Galatians 3:11

but My righteous one shall live by faith; and
if he shrinks back, My soul has no pleasure in
him.

Hebrews 10:38

For in it (the Gospel) the righteousness of
God is revealed from faith to faith; as it is
written, "but the righteous man shall live by
faith."

Romans 1:17

Which of these two universes is most likely to produce
people of the greatest faith? Throwing away the theory of
evolution and adopting instead a theory of intelligent design
eliminates the need for faith. Fundamental to intelligent
design is the claim that anyone can take an ordinary microscope and see for themselves genuine supernatural manifes-
tations which can only have come straight from the arm of
the Lord. There is no other revelation like this one, nor has
there ever been. No longer do Christians have to walk by
faith. Now they can walk by sight. If, however, faith in the
existence of God is still needed for the journey to the king-
dom of heaven, then intelligent design is not what it claims
to be – the physical evidence of the reality of God. Is contravening the Bible's many statements about the personal
significance of faith a clue as to which universe best fits
God's plans?

Continuing right along from Romans 1:17, the very next passages are often cited by the advocates of intelligent design as part of the scriptural bedrock supporting their contentions:

> For the wrath of God is revealed from heaven against all ungodliness and unrighteousness of men, who suppress the truth in unrighteousness, because that which is known about God is evident within them; for God made it evident to them. For since the creation of the world His invisible attributes, His eternal power and divine nature, have been clearly seen, being understood through what has been made, so that they are without excuse.
>
> *Romans 1:18-20*

The Epistle of Paul to the Romans, which is generally considered to have been written a few years before 60 A.D., has been referred to as "the cathedral of the Christian faith."[125] Here, at the beginning of this masterpiece, we are to understand from what is clearly seen around us, all of which God has made, that He exists. All creations have something to say about their creators and so nature also speaks of its Creator. A similar statement can be found in the Old Testament:

> The heavens are telling of the glory of God; and their expanse is declaring the work of His hands.
>
> *Psalms 19:1*

The proponents of intelligent design seem to prefer Romans 1:18-20 over Psalms 19:1, perhaps because it carries that condemnation of those whose activities are hostile to the truth, which in the context of Romans would mean, of course, the truth of the Bible. This overlooks the possibility that this truth may also include *all* the truth's of God's creation, such as those very many truths discovered by science yet never mentioned in the Bible, thereby exposing evolution-deniers as the ones "who suppress the truth in unrighteousness."

Whether you are fortunate enough to be a living witness to a miraculous revelation of God, or are simply gazing once again at the vast sweep of a beautiful starry night, that anything made by God would acknowledge God is a perfectly sensible view of our reality. God's deliberate interventions unmistakably disclose His truth, and those who have faith see clearly the work of His hands. Many people find good sense in this, including many scientists.[126] In fact, each year there are many people who would be more than glad to testify that they were led to Christ by the good sense of this.

In making and managing His creation, however, there are some things which God invariably does not do and one of them is to create within His creation a manifestation beyond nature that remains ever-present. Yet this is exactly what intelligent design claims via the so-called irreducible complexity of biological molecules. Nature, supposedly, cannot design these extraordinarily complex molecules any more than nature can drop manna from heaven, stop the sun and the moon in the sky, or raise the dead. Their complexity is therefore impossible, and thus miraculous. Unfortunately, this proposition is in conflict with biblical history. The long record consistently presents God revealing Himself in this fallen world by creating miracles whose reality is always ephemeral:

> Truly, Thou art a God who hides Himself
> *Isaiah 45:15*

Is contravening several thousand years of uniform biblical history a clue as to which universe best fits God's plans?

Perhaps evolution-deniers are on the wrong path, which is precisely what any modestly attentive individual would conclude just from the cacophony of their deceits about science. But if all this deceit sits well with you, then it should be relatively easy to take one step further and deny any damage intelligent design does to the meaning and purpose of the message delivered by the Bible. For those, however, who are not captives of evolution-denial, the universe of intelligent design isn't just far removed from science – it is a moral and scriptural disaster as well.

19

Take Up the Full Armor of Denial

Deceit is a word both famous and fabulous because it spans a range and depth of behavior so much greater than just plain and simple lying. With deceit you can tell lies even while speaking the truth, merely avoid revealing the entire truth. Although the full range of human intelligence, knowledge, skill, and creativity can be applied to works of deception, no amount of such effort has ever been known to change a single speck of the truth. Deceit, therefore, always constructs its edifices in the presence of the very truth which can destroy it. Like a house of cards, falsehoods are built from materials of imaginary strength and thus, the more extensive the deception, the more likely it is to fail. That Christian evolution-deniers apply their gifts to the construction of a "world view" built from deliberate fabrications and ignominious motivations, and by so applying themselves behave in ungodly ways, is frankly unmistakable. This brief work alone documents a driven and repetitive promotion of misinformation across an extensive range of knowledge, including:

- the fossil record

- the biology of moths

- the content of scientific papers

- mathematics

- biochemistry

- thermodynamics

- probability calculations

- the theory of evolution

- quotations from scientists

- scientific research

- the origins of life

- the character of Darwin

- even the history of the Holocaust

In sync with this sorry list are all the other corruptions portrayed on these pages, including especially a stunning, massive denial of any wrongdoing from a large number of individuals and organizations all linked together by the common claim of moral superiority. By fostering a mass movement of unchristian behavior, the leaders of evolution-denial subvert the Christian mission.

In a classic tragedy the main character achieves his own destruction through a series of horrendous decisions which he thinks are intelligent and correct. Throughout the journey he is repeatedly surprised by his seemingly inexplicable descent toward the final act of annihilation. Nevertheless, to the amazement of the audience, he foolishly "stays the course." Evolution-deniers are characters in a perfect tragedy. They stand in defiance against one of the most spectacular successes in science and maintain a dedicated cluelessness regarding the range and depth of their own transgressions. It is not for any want of a good reason that Christians have lost influence in the modern world. By deliberately undermining the facts of true science with their compendium of false science, the leaders of evolution-denial have led both themselves and their faithful followers down this self-destructive path.

The history of science is a history of progress. Year after year more and more evidence accumulates confirming the truth of evolution. The gaps in the fossil record consistently predict the discoveries of the future. Each year new discoveries fulfill old predictions, validating the theory and vindicating biologists again and again, all the while clarifying with ever diminishing gaps and ever increasing resolution the evolutionary history of life. All of this takes place to the accompaniment of other validation coming from numerous independent lines of investigation in both biological and non-biological sciences.[127]

Seeking to maintain their standing in their communities, the masters and minions of anti-evolution are driven to deny all this. So they take the turn down that tragic path and discover a wide open field of maneuver. They begin with a curl of genuine contempt on their lips while disparaging those "evilutionists" and then move on to orchestrating discord between Christians and non-believers, defining their enemies and vilifying them over and over, working diligently to make their own contributions to an extensive library of falsehoods, distortions and half-truths, designing their statements to deliberately keep "the lay audience" in ignorance, and end up complaining with real passion about imaginary fleas and insignificant mites in the hope that no one will notice the growing mammoth of evolution in the room. They twist and turn in every direction to deny every valid point, every true fact, while occasionally offering feigned respect for their enemies along with sporadic and inconsistent admissions of minor truths to complete a lame and stultifying disguise. Their purpose is to affirm their power and influence over their fellow Christians, which they can succeed at as long as they assert a believable self-assurance while keeping their followers misinformed and uninformed. Having started down this path and seeing no point in going halfway, they complete the journey with accusations that their opponents are engaged in the very transgressions they themselves repeatedly employ. The young are easily manipulated, as are the guileless, while the more thoughtful are encouraged by the local group-think to avoid any disturbances which honest discernment might derive. Expounding from their false writ both passionately and desperately, the instructors of anti-evolution and intelligent design are propelled across this iniquitous landscape by a need to assuage their anxiety over facts their narrow vision perceives as intolerable, and to that end they have foolishly constrained the infinite God in a little box of unimaginative and controlled behavior, constructed from words and images which are constantly under repair with the glue of their own sins.

The fundamental spiritual error made by Christian evolution-deniers is that they fear man. Clearly, they do not fear God:

> Transgression speaks to the ungodly within his heart; there is no fear of God before his eyes.
>
> *Psalms 36:1*

> The fear of man brings a snare, but he who trusts in the Lord will be exalted.
>
> *Proverbs 29:25*

When have the perpetrators of deceit not been a part of our lives? They are always around in every society, in every large organization, in every profession, and in every moment of history – sometimes they can even be found in a family. There will always be people who think that success comes from telling lies, manipulating and misleading, creating conflicts and nurturing them, spreading misinformation. Since they always exist, the real problem is how many followers they can control with their deceptions. The Christian evolution-denial movement has a lot of followers. Their number is a measure of the damage done to Christianity. If you are one of these people, by following this path of deceit and bending with the denial of it, you have made your own tragic contribution to the decay and debris of Christian influence.

20

Sapphira

The story of Sapphira plays out at the beginning of Acts 5. In the earliest days of the Christian church, literally in its first months, some enthusiastic believers sold their property and placed the money at the feet of the Apostles to be distributed as need arose. A man named Ananias, with his wife Sapphira, decided to join in this generosity – sort of. Ananias sold some property but told his wife he was going to keep part of the proceeds for himself. When he dropped off the money, Peter, who was on fire with the Holy Spirit, confronted him with the fact that he had secretly held back some for himself and pointed out that it was not men he had lied to, but God. Ananias dropped dead on the spot.

A few hours later Sapphira showed up, unaware that her husband was dead. Peter asked her if the money rendered was the full amount and she replied that it was. When he then exposed her part in the fraud, " . . . she fell immediately at his feet, and breathed her last."[128]

Obviously Ananias was the driving force of this deception while his wife Sapphira was merely a follower. In point of fact her position was a bit more complex than that of just a follower. As a dutiful wife immersed in a male-dominate culture, she was subject to numerous influences directing her obedience to her husband. Her lie was thereby the easiest to tell. She was just going with the flow. As generations of Bible students have observed however, she did have plenty of time to reflect on her choices.

In some far off corner of the country there may be someone who still doesn't know that the intelligent design movement appeared on the scene as a disguised Christian agenda with the intent of slipping the Christian God past the Supreme Court and around the Constitution of the United States and back into the public schools. Here is Philip E. Johnson laying it out:

> Our strategy has been to change the subject a bit so that we can get the issue of intelligent design, which really means the reality of God, before the academic world and into the schools.[129]

The "Intelligent Designer" is a mask put over God. It is a subterfuge, a ruse, a truth concealed. From some source of affirmation the leaders of Christian evolution-denial claim the authority to tell and repeat lies about science and about scientists, even going so far as to cover God in public with a pseudo-science disguise. Yet throughout the New Testament, both during the life of Jesus and afterwards, there is no precedent for disguising the truth, or "to change the subject a bit." The legacy, and therefore the message, is just the opposite – a complete openness in telling the truth, even under the threat of certain death.

The followers of Christian evolution-denial are not bound to their leaders as Sapphira was bound to her husband. If the bond between a husband and his wife is no excuse to God for any kind of sin, then what excuse do these followers have – an excusable ignorance because science is too hard to understand, an excusable lack of discernment because all Christians are to be trusted by default, an excusable motive to show solidarity with Team Christ?

As the evidence of flourishing sin generated by the active proponents of anti-evolution and intelligent design continues to accumulate with new books, new videos, and daily updates to numerous, dishonest web pages, are all their followers merely a migration of mindless lemmings racing across green fields, just going with the flow? Are they nothing more than a mesmerized legion of lost sheep led astray by a handful of stiff-necked manipulators? More likely the vast majority are ordinary people busy each day with the tasks of raising families and paying bills and on top of all that, hard as it may be for some to believe, they actually have never had any real interest in science. They are neither "mindless" nor "mesmerized." In their voluntary development as Christians, in their studies of the Bible with its messages and promises, it has simply never occurred to them that the Christian evolution-denial movement is something they might need to be particularly discerning about. For each individual discerning the truth can often become a long running drama. It is not always a one-off event acquired from some page in a book. It can be a journey of character and courage.

21

The Sand Castle

Sand, though made of rock, holds only the shape of fantasies. In building up their world view, evolution-deniers have constructed their edifice out of sand. While claiming to speak the truth, they in fact stand on a foundation of half-truths, outright lies and intellectual trickery. While they pretend enlightenment and claim to encourage critical thinking, they in fact promote ignorance, insist on an exaggeration of partial explanations, and cultivate a reduction in ordinary rational thought. While they display the trappings of authority, their authority is in fact a sham constructed out of an image of God that God Himself does not portray. The three principal failings of Christian evolution-denial are:

- Scientifically – not only is there no physical evidence which supports their ideas, and not only does all of the physical evidence support the idea they reject, but there is also every reason to expect that the evidence against their ideas will continue to accumulate as it has done for the last one hundred and fifty years.

- Scripturally – the claim that biological molecules reveal a supernatural complexity is not an example of any known behavior of God. The claim is contrary to the ephemeral character of every supernatural manifestation ever recorded, testified to, or personally experienced. Furthermore, proof of God cannot be found as a commonplace element of nature without cancelling the fundamental separation between God and man and thereby undermining all biblical pronouncements regarding God's desire for people to live by faith.

- Spiritually – anti-evolution and intelligent design are widely supported and well funded generators of sin in the Christian community and the constant denial of this sin is the definition of hypocrisy.

Apparently intending to fool the world, evolution-deniers succeed only in fooling themselves. With the abundance, conspicuousness, and depth of their errors, along with their dedication to them, perhaps fooling themselves has been the real goal all along. If human nature can serve as a guide, this fraudulent structure won't be torn down by the people who built it and maintain it, but by the people being told to live in it.

Someday a new group of Christians will come along who are unafraid of the full truth of life in this extravagant universe, who are cognizant of the damage done to themselves and to Christianity by this long exhibition of flagrant fakery, and whose reaction to this betrayal of their values will be plain honesty. It is these people who will pull the Christian movement out of this dark dead-end and give it a new relevance in the modern world. Instead of cloistering themselves in a contrived, private Christian artifice, they will reach out into the world by, among other things, illuminating the obvious link between the infinite God and all of the discoveries of science, including the theory of evolution in all its scientific exactitude.

22

A Biblical Life in a Scientific Civilization

If your eyes are sharp and the night is clear and dark, you can see in the heavens near the tip of Scorpio's stinger, at the boundary with Sagittarius, a faint fuzzy glow. It was first mentioned by Ptolemy around 130 A.D., although he almost certainly relied on Hipparchus from three hundred years earlier. With the invention of the telescope the fuzzy blot was finally revealed as a small cluster of bright blue stars. By 1654 the nebula hunter Giovanni Hodierna had counted thirty individuals. In the late spring of 1764 Charles Messier, compiling a list of hazy objects which might fool comet hunters, added the fuze ball as the seventh entry in his catalogue. Today the little star cluster is known as Messier 7, or simply M7.

One of the more remarkable features in practically every modern image of M7 is not M7 at all, but what lies in the background. Splayed across the image like grains of sand are literally millions of distant stars. The star cluster looks like a handful of jewels dropped on a beach.[130] Most of the multitude of background stars are, like our sun, billions of years old. If the sun, which is almost nine hundred thousand miles across, were added into the background of M7, it would not be noticed. On the full scale of the cosmos this entire scene is just a dust mote on a dust mote, and even that vastly understates our physical insignificance.

The subatomic particles and high energy photons blasting out of radioactive material demonstrate quite convincingly that the universe has a randomness all the way to its core. This characteristic is also apparent in snowflakes, clusters of galaxies, natural disasters, any roll of the dice, and in the fossil record. Within the bounds of science each one of us is accurately described as an insignificant random creation of nature living in a gigantic universe which has no apparent purpose.

To the Christian God, however, each one of us is no such thing:

> just as He chose us in Him before the foundation of the world, that we would be holy and blameless before Him.
>
> *Ephesians 1:4*

To be chosen by God "before the foundation of the world" is the highest possible accolade. But it does place us at odds with scientific fact. This discrepancy between religion and science can be resolved either by (i) denying God, as some do, or (ii) denying science, as we have seen in this brief work, or, if either of these options seem unwise, then (iii) acknowledging that any disparity is meaningless when standing in

145

the presence of God Almighty, the God of gods, the Everlasting God, the LORD, the Creator of the ends of the earth, the Ancient of Days, the first and the last.[131]

It is a well known fact that the arm of the Lord has a reach and a capability ultimately indiscernible by any human intellect and therefore unspecifiable in any language of man.

> "God thunders with His voice wondrously, doing great things which we cannot comprehend."
>
> *Job 37:5*

> Oh, the depth of the riches both of the wisdom and knowledge of God! How unsearchable are His judgments and unfathomable His ways! For who has known the mind of the Lord, or who became His counselor?
>
> *Romans 11:33-34*

> "For My thoughts are not your thoughts, neither are your ways My ways," declares the Lord.
>
> *Isaiah 55:8*

Since your mind has no infinite capabilities, as long as you are in this world you are never going to comprehend the full truth of God – you cannot accurately define Him, you cannot successfully predict Him, you cannot entirely explain Him, and you have no way of acquiring with certainty a complete understanding of the Bible unless you wish to take the extraordinary step of claiming final knowledge of God. All the accuracies of science are thereby accommodated as

146

expressions of the unlimited God. In particular, to God, who apparently chose you before it all started, the scientific randomness of nature creates precisely what He already knows.

Because science is a process of uncovering the truth about nature, there can be no incompatibility between the truth of the infinite God, who created nature, and anything science verifies, including the theory of evolution. The incompatibility which the Bible addresses on every single page is between man and God. The many components of Christian anti-evolution assemble into a juggernaut of sin which is plowing through the Christian community, speaking clearly of this incompatibility while damaging and toppling the good works of many others.

The instructors of intelligent design and anti-evolution deny that their structure is built from contradictions and transgressions. But the fundamental intelligent design claim that evidence of actions beyond nature reside in our presence brings man into contact with God which is contrary to the separation of man and God, which is fundamental to Christianity. And although Christianity promotes itself as the very heart of both truth and trust, the promoters of anti-evolution have a long history of molding obvious distortions into products they sell to the deceived. While the truths of science expose them and their own iniquities diminish them, they nevertheless succeed in sustaining their cause by repeatedly enlisting the powers of denial.

If you think Christian evolution-deniers are not telling rampant lies about science, you are in denial. If you are a Christian and you think there are no consequences for supporting and propagating lies about science, you are in denial. Within the rationale of Christian anti-evolution and intelligent design, God is a simpleton who can't even establish laws of nature that produce an all-natural complexity generator for biological molecules. And while attempting to construct a separation between Himself and mankind, consigning humanity to a "fallen world," the very essence of which is a natural, godless explanation for *everything*, in point of fact He has failed at this task, instead filling the universe everywhere with supposedly obvious evidence of His designs – or so they say. There are no ungodly ways more destructive than those found in the lawlessness which accompanies the worship of a reduced God. But in the vast and complex reality defined by our scientific discoveries, including the theory of evolution, the only God who could create such a universe is a God of unlimited sophistication.

Reference Notes

1. Philip E. Johnson, *Darwin on Trial* (Downers Grove: InterVarsity Press, 2nd edition, 1993) 48.

2. Charles R. Darwin, *On the Origin of Species by Means of Natural Selection, or the Preservation of Favoured Races in the Struggle for Life* (London: John Murray, 1859) 342.

    Title changed to *The Origin of Species . . .* with the 6th edition, 1872.

    Available at:
    The Complete Works of Charles Darwin Online – On the Origin of Species,
    http://darwin-online.org.uk/contents.html#origin

3. Stephen Jay Gould, "Impeaching a Self-Appointed Judge," *Scientific American*, July 1992: 118-21.

    Available at:
    http://www.stephenjaygould.org/ctrl/gould_darwin-on-trial.html

4. Johnson, *Darwin on Trial*, 1.

5. Eugenie C. Scott, "Darwin Prosecuted: Review of Johnson's Darwin on Trial," *Creation Evolution Journal* Vol. 13. No. 2, Winter 1993.

    Available at:
    http://ncseweb.org/cej/13/2/darwin-prosecuted-review-johnsons-darwin-trial

    Mark Perakh, "A Militant Dilettante in Judgement of Science, How a lawyer (Phillip Johnson) disproves Darwinism," Talk Reason, 2000, http://www.talkreason.org/articles/johnson.cfm

6. Johnson, *Darwin on Trial*, 160.

7.   Colin Patterson, *Evolution* (Ithaca: Cornell University Press, 2nd edition, 1999) 55.

8.   Wikipedia, "Peppered moth evolution," http://en.wikipedia.org/wiki/Peppered_moth_evolution

9.   E-mail from M. E. N. Majerus to Donald Frack, April 5, 1999:

     "Evidence of selective predation in the peppered moth is not lacking. It is just not provided in the quick text book descriptions of the peppered moth. How can it be. I have read some 500 papers on melanism in the Lepidoptera. In total, these papers probably amount to about 8000 pages, ad(sic) the story is condensed into a few paragraphs in most textbooks for schools. Even in my own book, I could only give a review of the case covering about 60 pages including illustrations."

     Available at:
     "Peppered Moths - round 2 (part 2 of 2)," http://www.asa3.org/archive/evolution/199904/0103.html

10. Steve Connor, "Moth study backs classic 'test case' for Darwin's theory," The Independent, August 25, 2007.

     Available at:
     Freethoughtpedia, "Peppered Moth," http://freethoughtpedia.com/wiki/Peppered_Moth

11. True Parents Organization, "Darwinism: Why I Went for a Second Ph.D.," http://www.tparents.org/library/unification/talks/wells/DARWIN.htm

     Wikipedia, "Jonathan Wells (intelligent design advocate)," http://en.wikipedia.org/wiki/Jonathan_Wells_(intelligent_design_advocate)

12. Jonathan Wells, *Icons of Evolution: Science or Myth? Why Much of What We Teach About Evolution Is Wrong* (Washington, DC: Regnery Publishing, 2000) 139.

13. *Ibid.* 150.

14. Michael E. N. Majerus, *Melanism: Evolution in Action* (Oxford: Oxford University Press, 1998) 123.

15. *Ibid.* 116.

16. Wikipedia, "Peppered moth evolution."

17. Michael E. N. Majerus, *Moths* (London: HarperCollins Publishers, 2002) 252.

18. Wells, *Icons of Evolution*, 143.

19. Claim CB601: "The traditional peppered moth story is no longer supportable," ed. Mark Isaak, The TalkOrigins Archive, Index to Creationist Claims, 2005, http://www.talkorigins.org/indexcc/CB/CB601.html

     See also:
     Tommy Mitchell, "Much Ado About Moths," Answers in Genesis, April 23, 2008, http://www.answersingenesis.org/articles/aid/v3/n1/much-ado-abo ut-moths

20. Darwin, *The Origin of Species*, 469.

21. Alan D. Gishlick, "Icons of Evolution? Why much of what Jonathan Wells writes about evolution is wrong," National Center for Science Education, http://ncseweb.org/creationism/analysis/icons-evolution

     Nick Matzke, "Icon of Obfuscation," The TalkOrigins Archive, 2002, http://www.talkorigins.org/faqs/wells/iconob.html

22. Stephen C. Meyer, "Teach the Controversy on Origins," *Cincinnati Enquirer*, Commentary Section, March 30, 2002.

     Available at:
     http://www.arn.org/docs/meyer/sm_teachthecontroversy.htm

23. NCSE Staff, "Analysis of the Discovery Institute's Bibliography," National Center for Science Education, http://ncseweb.org/creationism/general/analysis-discovery-institute s-bibliography

24. *Ibid.*

25. William A. Dembski, *The Design Inference: Eliminating Chance Through Small Probabilities* (Cambridge: Cambridge University Press, 1998) 36.

26. William A. Dembski, *Intelligent Design: The Bridge Between Science & Theology* (Downers Grove: InterVarsity Press, 1999) 134.

   See also:
   Mark Frank, "Detecting design: Specification versus Likelihood," Talk Reason, July 2, 2006,
   http://www.talkreason.org/articles/likely.cfm

27. Wikipedia, "Pulsar,"
   http://en.wikipedia.org/wiki/Pulsar#Discovery

28. Matt Young, "Dembski's Explanatory Filter Delivers a False Positive," The Panda's Thumb, April 22, 2004,
   http://pandasthumb.org/archives/2004/04/dembskis-explan.html

   See also:
   John Sweeney and Bill Law, "Gene find casts doubt on double 'cot death' murders," The Observer, July 15, 2001,
   http://www.guardian.co.uk/uk/2001/jul/15/johnsweeney.theobserver

29. Michael J. Behe, *Darwin's Black Box: The Biochemical Challenge to Evolution* (New York: Touchstone, 1996).

30. Hermann J. Muller, "Genetic variability, twin hybrids and constant hybrids, in a case of balanced lethal factors," *Genetics* 3, 1918: 464,
   http://www.genetics.org/cgi/reprint/3/5/422

   See also:
   Chris Ho-Stuart, "Irreducible Complexity as an Evolutionary Prediction," The TalkOrigins Archive, Post of the Month, September 2006,
   http://www.talkorigins.org/origins/postmonth/sep06.html

   Douglas Theobald, Ph.D., "The Mullerian Two-Step: Add a part, make it necessary or, Why Behe's 'Irreducible Complexity' is silly," The TalkOrigins Archive, July 18, 2007,
   http://www.talkorigins.org/faqs/comdesc/ICsilly.html

31. Alex Fidelibus, "The Evolution of the Mousetrap through Natural Selection,"
   http://www.fidelibus.com/mousetrap

   John H. McDonald, "A Reducibly Complex Mousetrap," 2002,
   http://udel.edu/~mcdonald/mousetrap.html

32. Kenneth R. Miller, "A Review of *Darwin's Black Box,*" *Creation Evolution Journal* 16, 1996: 36-40.

Available at:
http://www.millerandlevine.com/km/evol/behe-review/index.html

33. Russell F. Doolittle, "A Delicate Balance," *Boston Review,* February/March 1997,
http://bostonreview.net/BR22.1/doolittle.html

34. Kitzmiller v. Dover Area School Dist., 400 F.Supp.2d 707, 741 (M.D.Pa. 2005).

Available at:
The U. S. District Court for the Middle District of Pennsylvania, "Memorandum Opinion December 20, 2005:" 78,
http://www.pamd.uscourts.gov/kitzmiller/kitzmiller_342.pdf

35. Henry Morris, *The Remarkable Birth of Planet Earth* (Bethany Fellowship, Inc., Minneapolis, 1972) 14.

36. David A. Noebel, *Understanding the Times: The Religious Worldviews of Our Day and the Search for Truth* (Harvest House Publishers, 1st Edition, 1991) 331.

37. Wikipedia, "Second law of thermodynamics,"
http://en.wikipedia.org/wiki/Second_law_of_thermodynamics

38. Since there are thirty characters in each sequence, and each character could be any of twenty seven possibilities – all of the capital letters of the alphabet plus the period, the total number of different sequences is twenty seven to the thirtieth, which is twenty seven times itself thirty times (27 x 27 x 27 x . . .), which equals, approximately, eighty seven followed by forty one zeros.

39. Determinations of the life span of the universe are, as you would expect, fraught with uncertainty due to our ignorance as to the full nature of the cosmos. Nevertheless, there are usually a few ways to make an intelligent estimate. One possible answer comes from a paper (see below) which reports the failure of orbiting telescopes to detect any clear indication of x-rays emitted by the decay of dark matter particles in the collisions of clusters of galaxies. These enormous encounters tend to separate huge amounts of dark matter from ordinary visible matter. The lack of decay radiation from these unique regions of space suggests that the lifetime of dark matter particles may

be greater than 3,000,000 billion years. Since the gravitational effects of dark matter demonstrate that it makes up by far most of the material in the universe, we can use this estimate for the lifetime of these mysterious particles as an approximation for the life span of the entire universe.

Running through all possible thirty character sequences (eighty seven followed by forty one zeros) at a speed of a million every second would take about 2.76 times ten to the twenty-ninth years. If the life span of the universe is taken from the above estimate for the lifetime of dark matter particles, then it would take about 92,000 *billion* lifetimes of the universe to go through all of the different character sequences.

Signe Riemer-Sørensen, et al., "Searching for Decaying Axionlike Dark Matter from Clusters of Galaxies," *Physical Review Letters*, Vol. 99, Issue 13, 131301, September 2007: 1-4, http://link.aps.org/doi/10.1103/PhysRevLett.99.131301

See also:
University of Copenhagen, "Dark Matter Of The Universe Has A Long Lifetime," *ScienceDaily*, 4 October 2007, http://www.sciencedaily.com/releases/2007/10/071001112906.htm

40. Anthony I. Okoh et al., "Aerobic Heterotrophic Bacterial and Fungal Communities in the Topsoil of Omo Biosphere Reserve in Southwestern Nigeria," *Biotropica*, Vol. 32, Issue 2, June 2000: 208-12.

Available at:
http://onlinelibrary.wiley.com/doi/10.1111/j.1744-7429.2000.tb00463.x/abstract

41. Jung-Ho Hyun and Eun-Jin Yang, "Freezing Seawater for the Long-term Storage of Bacterial Cells for Microscopic Enumeration," *The Journal of Microbiology*, Vol. 41, No. 3, September 2003: 262-5.

Available at:
http://www.msk.or.kr/jsp/downloadPDF1.jsp?fileName=413-15.pdf

42. Erwan G. Roussel et al., "Extending the Sub-Sea-Floor Biosphere," *Science* Vol. 320, No. 5879, 23 May 2008: 1046, http://www.sciencemag.org/content/320/5879/1046.abstract

43.   Patterson, *Evolution*, 30.

See also:
Stanley L. Schrier, MD and Stephen A. Landaw, MD, PhD., "Introduction to hemoglobin mutations," UpToDate, http://www.uptodate.com/patients/content/topic.do?topicKey=~Or EJEbFF8hpe85O

44.   Darwin, *The Origin of Species*, 116-7.

The Diagram is also available at:
The Friends of Charles Darwin, "Diagram from the Origin of Species," http://friendsofdarwin.com/docs/origin-1/diagram

45.   Wikipedia, "Mexican tetra," http://en.wikipedia.org/wiki/Mexican_tetra

46.   D. R. Maddison and K.S. Schulz (Editors), The Tree of Life Web Project, 2007, http://tolweb.org/tree

47.   Darwin, *The Origin of Species*, 84.

48.   Fazale R. Rana, "Tetrapod Transitions: Evidence for Design," *New Reasons To Believe* Vol. 1, No. 1, 2009: 6-7,

Available at:
http://www.reasons.org/articles/tetrapod-transitions-evidence-for-d esign

49.   All About Creation, "Evidence For Evolution," http://www.allaboutcreation.org/evidence-for-evolution.htm

Creation Ministries International, "That quote! — about the missing transitional fossils," http://creation.com/that-quoteabout-the-missing-transitional-fossils

The Bible Probe, "Objections to the Doctrine of Evolution," http://bibleprobe.org/objection.html

50.   Noebel, *Understanding the Times*, 279.

James Porter Moreland (Editor), *The Creation Hypothesis: Scientific Evidence for an Intelligent Designer* (InterVarsity Press, Downers Grove, IL, 1994) 280.

REFERENCE NOTES

James Perloff, *Tornado in a Junkyard: The Relentless Myth of Darwinism* (Refuge Books, Arlington, MA, 1999) 11.

Philip Stott, *Vital Questions* (Reformation Media & Press, 2nd edition, 2002) 85.

Donald W. Ekstrand, *Christianity: The Pursuit of Divine Truth* (Xulon Press, 2008) 52.

51. YouTube, *Transitional Fossils? Evolutionists Say No!* at 8 minutes, 34 seconds, http://www.youtube.com/watch?v=typrxYyiEEI

52. Del Tackett, *The Truth Project DVD Set*, (Colorado Springs: Focus On The Family, 2006): Lesson 5b, Slide #16.

53. Wikipedia, *"Australopithecus afarensis,"* http://en.wikipedia.org/wiki/Australopithecus_afarensis

54. University of California Museum of Paleontology, *"Archaeopteryx*: An Early Bird,"* http://www.ucmp.berkeley.edu/diapsids/birds/archaeopteryx.html

55. Understanding Evolution, "What has the head of a crocodile and the gills of a fish?" (May 2006), http://evolution.berkeley.edu/evolibrary/news/060501_tiktaalik

56. Wikipedia, "Emperor of Japan," http://en.wikipedia.org/wiki/Emperor_of_Japan

57. Wikipedia, "Ursid hybrid," http://en.wikipedia.org/wiki/Ursid_hybrid

58. Claim CB910: "New Species," ed. Mark Isaak, The TalkOrigins Archive, Index to Creationist Claims, 2007, http://www.talkorigins.org/indexcc/CB/CB910.html

59. *Ensatina* web site, *"Ensatina eschscholtzi*, Speciation in Progress,"* http://www.ensatina.net

60. Wikipedia, "Evolution of cetaceans," http://en.wikipedia.org/wiki/Evolution_of_cetaceans

61. Patterson, *Evolution*, 109.

62. Lionel Theunissen, "Patterson Misquoted: A Tale of Two 'Cites'," The
     TalkOrigins Archive, 1997,
     http://www.talkorigins.org/faqs/patterson.html

63. "Peace, peace to him who is far and to him who is near," says the
     Lord, "and I will heal him." But the wicked are like the tossing sea, for
     it cannot be quiet, and its waters toss up refuse and mud. "There is no
     peace," says my God, "for the wicked."

                                                            Isaiah 57:19-21

64. Greta Keller et al., "New evidence concerning the age and biotic
     effects of the Chicxulub impact in NE Mexico," *Journal of the Geological
     Society* Vol. 166, Part 3, May 2009: 393-411,
     http://jgs.geoscienceworld.org/cgi/content/abstract/166/3/393

     Peter Schulte et al., "The Chicxulub Asteroid Impact and Mass
     Extinction at the Cretaceous-Paleogene Boundary," *Science*, Vol. 327.
     no. 5970, March 5, 2010: 1214-18,
     http://www.sciencemag.org/content/327/5970/1214.abstract

65. Seulki Kim, "Conviction of disgraced SKorean scientist upheld," *The
     Associated Press*, 2010.

     Available at:
     Fox News, December 16, 2010
     http://www.foxnews.com/world/2010/12/16/conviction-disgraced-s
     korean-scientist-upheld

66. Darwin, *The Origin of Species*, 1.

67. *Ibid.* 67-8.

68. *Ibid.* 93.

69. *Ibid.* 362.

70. *Ibid.* 25, 71, 98, 114, 135, 148, 183, 220, 228, 358, 386, and 445.

71. *Kitzmiller v. Dover Area School District*, "Trial transcript: Day 12
     (October 19, 2005), AM Session, Part 2", The TalkOrigins Archive,
     http://www.talkorigins.org/faqs/dover/day12am2.html

72. *Ibid.*, "Trial transcript: Day 20 (November 3), PM Session, Part 2", The
     TalkOrigins Archive,
     http://www.talkorigins.org/faqs/dover/day20pm2.html

73. *Ibid.*, "Trial transcript: Day 21 (November 4), AM Session, Part 2", The TalkOrigins Archive, http://www.talkorigins.org/faqs/dover/day21am2.html

74. "Bacterial Generation time and example of fast growing bacteria and slow growing bacteria?" WikiAnswers, http://wiki.answers.com/Q/Bacterial_Generation_time_and_exampl e_of_fast_growing_bacteria_and_slow_growing_bacteria

Under ideal conditions a bacteria such as E. Coli can produce a new generation about every twenty minutes. To run through 10,000 generations would then take about five months. But if this estimate is too low and it takes ten times as many generations for the evolution of a flagellum, then at an average generation time of twenty minutes, 100,000 generations would take approximately four years. Usually many runs of an experiment are required to get the techniques and conditions right before any repeatable results can be expected to appear in a published paper. What if the bacteria required a million generations to evolve a flagellum? To run such an experiment just once would then take 38 years. But what if ten million generations were required?

75. Adams, James W. (Producer) & Allen, Lad (Director), *Unlocking the Mysteries of Life* [Video], United States: Illustra Media, 2003.

Available at:
http://www.youtube.com/watch?v=VWvS1UfXl8k

76. Wikipedia, "Scientific method," Characterizations, Definition, http://en.wikipedia.org/wiki/Scientific_method#Characterizations

77. Wikipedia, "Life," http://en.wikipedia.org/wiki/Life#Definitions

78. Fred Hoyle and N. C. Wickramasinghe, *Evolution From Space: A Theory of Cosmic Creation*, (New York: Simon & Schuster, 1981) 24.

79. David Foster, *The Philosophical Scientists*, (New York: Barnes & Noble, 1993) 82.

80. Richard Carrier, "Addendum B: Are the Odds Against the Origin of Life Too Great to Accept?," The Secular Web, http://www.infidels.org/library/modern/richard_carrier/addendaB. html

159

81. Fred Hoyle, *The Intelligent Universe: A New View of Creation and Evolution*, (New York: Holt, Rinehart and Winston, 1984) 19.

82. Orville Wright, *How We Invented The Airplane*, (New York: Dover Publications, 1988) 85-6.

    See also:
    Wikipedia, "Wright Brothers,"
    http://en.wikipedia.org/wiki/Wright_brothers

83. Wikipedia, "Natural nuclear fission reactor,"
    http://en.wikipedia.org/wiki/Natural_nuclear_fission_reactor

84. Charles R. Darwin, *Voyages of the Adventure and Beagle, Vol. III.* (London: Henry Colburn, 1839) 28.

    Available at:
    The Complete Works of Charles Darwin Online, Journal of Researches (or Voyage of the Beagle), "Darwin. 1839. *Journal and remarks. 1832-1836.*"
    http://darwin-online.org.uk/contents.html#researches

85. Barbara Forrest and Paul R. Gross, *Creationism's Trojan Horse, The Wedge of Intelligent Design*, (New York: Oxford University Press, 2005) 221.

86. *Tammy Kitzmiller, et al v. Dover Area School District, et al.* 400F. Supp. 2d 707 (M.D. Pa. 2005) 137-8.

87. Staff, "'Masterful' Federal Ruling on Intelligent Design Was Copied from ACLU," Discovery Institute, December 12, 2006,
    http://www.discovery.org/a/3828

88. David Hardy, "Michael Moore Exposed,"
    http://www.mooreexposed.com

    See also:
    Wikipedia, "Fahrenheit 9/11,"
    http://en.wikipedia.org/wiki/Fahrenheit_9/11

89. Eugenie C. Scott, "A Rude Introduction to "Expelled"," *Reports of the National Center for Science Education*, Vol. 28. Issue 5-6, (September-December, 2008): 11-2,
    http://ncseweb.org/rncse/28/5-6/rude-introduction-to-expelled

90. "A Conversation with *Expelled's* Associate Producer Mark Mathis", *Scientific American*, April 9, 2008, at 18 minutes, 54 seconds, http://www.scientificamerican.com/article.cfm?id=a-conversation-w ith-mark-mathis

91. The Clergy Letter Project, "Christian Clergy Letter," http://theclergyletterproject.org

92. Ruloff, W. (Producer), & Frankowski, N. (Director), *Expelled: No Intelligence Allowed* [Motion picture], United States: Premise Media Corporation, 2008.

93. NCSE's staff, Expelled Exposed: The Truth behind the Fiction, National Center for Science Education, "Richard Sternberg," http://www.expelledexposed.com/index.php/the-truth/sternberg

94. *Ibid.,* "Guillermo Gonzalez," http://www.expelledexposed.com/index.php/the-truth/gonzalez

95. Ruloff, W., *Expelled: No Intelligence Allowed*: minute 53.

96. *Ibid.,* minute 66.

97. Wikipedia, "Godfrey of Bouillon," http://en.wikipedia.org/wiki/Godfrey_of_Bouillon

    See also:
    Cecil Roth and Geoffrey Wigoder, eds., "CRUSADES, The First Crusade," *Encyclopaedia Judaica*, (Jerusalem: Keter Publishing House Ltd., 1972) **5**, 1135.

    Available at:
    The Jewish Encyclopedia, The Crusades, "First Crusade: 1096," http://www.jewishencyclopedia.com/view.jsp?artid=908&letter=C& search=Crusades

98. David M. Dickerson, "Clifford's Tower: Massacre at York (1190)," July 2, 1997, http://ddickerson.igc.org/cliffords-tower.html

    See also:
    Paul Halsall, "Medieval Sourcebook: Ephraim of Bonn: The York Massacre 1189-90," February 1996, http://www.fordham.edu/halsall/source/ephr-bonn1.asp

99. New Advent – The Catholic Encyclopedia, "Fourth Lateran Council
(1215)," 2009,
http://www.newadvent.org/cathen/09018a.htm

See also:
Steven Kreis, "Lecture 29 – Satan Triumphant: The Black Death," The
History Guide: Lectures on Ancient and Medieval European History,
August 03, 2009,
http://www.historyguide.org/ancient/lecture29b.html

100. Roth, "ENGLAND, The Expulsion," *Encyclopaedia Judaica*, **6**, 751.

Available at:
The Jewish Encyclopedia, "England," The Expulsion,
http://www.jewishencyclopedia.com/view.jsp?artid=375&letter=E&
search=England#957

101. Roth, "ECONOMIC HISTORY, Medieval Christendom," *Encyclopaedia
Judaica*, **16**, 1284.

Available at:
The Jewish Encyclopedia, "France," Exile of 1306,
http://www.jewishencyclopedia.com/view.jsp?artid=288&letter=F&
search=France#820

102. Roth, "BLACK DEATH," *Encyclopaedia Judaica*, **4**, 1063-7.
Ibid., "BASLE (Basel)," **4**, 303.
Ibid., "STRASBOURG," **15**, 423.

Available at:
The Jewish Encyclopedia, "Black Death,"
http://www.jewishencyclopedia.com/view.jsp?artid=1114&letter=B
&search=BlackDeath

Ibid., "Basel," The Black Death,
http://www.jewishencyclopedia.com/view.jsp?artid=375&letter=B&
search=Basel

The History of the Jewish People, "1348 - 1349 THE BLACK PLAGUE
(Europe),"
http://www.jewishhistory.org.il/history.php?startyear=1340&endye
ar=1349

Paul Halsall, Jewish History Sourcebook, "The Black Death and the

Jews 1348-1349 CE,"
http://www.fordham.edu/halsall/jewish/1348-jewsblackdeath.html

Wikipedia, "Basel massacre,"
http://en.wikipedia.org/wiki/Basel_massacre

Switzerland Is Yours, "Basel: some history,"
http://switzerland.isyours.com/e/guide/basel/history.html

Steven Kreis, "Lecture 29 – Satan Triumphant: The Black Death,"
http://www.historyguide.org/ancient/lecture29b.html

103.   Martin Luther, *On the Jews and Their Lies*, (Fortress Press & Augsburg
       Fortress, 1971).

       Available at:
       http://www.humanitas-international.org/showcase/chronography/d
       ocuments/luther-jews.htm

104.   This Day In Jewish History, "1753 May 26, ZHITOMIR (Russia),"
       http://www.jewishhistory.org.il/history.php?startyear=1750&endye
       ar=1759

       See also:
       European History Online, "Ashkenazi Jews in Early Modern Europe,"
       Note #34,
       http://www.ieg-ego.eu/en/threads/europe-on-the-road/jewish-migr
       ation/ashkenazi-jews-in-early-modern-europe/predrag-bukovec-as
       hkenazi-jews-in-early-modern-europe

       Simon M. Dubnow, *History of the Jews in Russia and Poland*, Trans. I.
       Friedlander, (Philadelphia: The Jewish Publication Society of America,
       1918) 86.

105.   Adolf Hitler, *Mein Kampf* (Boston: Houghton Mifflin, 1999) 65.

       A slightly different translation is available at:
       Mein Kampf (1941), Internet Archive, (New York: Reynal And
       Hitchcock, 1941) 84.
       http://archive.org/stream/meinkampf035176mbp#page/n125/mode/
       2up

106.   Ruloff, W., *Expelled: No Intelligence Allowed*: minute 78.

107.   Charles R. Darwin, *The Descent of Man, and Selection in Relation to Sex* (London: John Murray, 1871) 168-9.

        Available at:
        The Complete Works of Charles Darwin Online, "The Descent of Man,"
        http://darwin-online.org.uk/contents.html#descent

108.   Wikipedia, "Jim Jones,"
        http://en.wikipedia.org/wiki/Jim_Jones

109.   Adolf Hitler, *Mein Kampf*, 231-2.

        A slightly different translation is available at:
        Mein Kampf (1941), Internet Archive, (New York: Reynal And Hitchcock, 1941) 313.
        http://archive.org/stream/meinkampf035176mbp#page/n353/mode/2up

110.   Ruloff, W., *Expelled: No Intelligence Allowed*: minute 29.

111.   Wikipedia, "Prediction," Prediction in Science,
        http://en.wikipedia.org/wiki/Prediction_in_science#Prediction_in_science

112.   Darwin, *The Origin of Species*, 341-2.

113.   Chris Nedin, "All About *Archaeopteryx*," The TalkOrigins Archive, 1999,
        http://www.talkorigins.org/faqs/archaeopteryx/info.html

114.   Wikipedia, "Origin of birds," Modern research and feathered dinosaurs in China,
        http://en.wikipedia.org/wiki/Dinosaur-bird_connection#Modern_research_and_feathered_dinosaurs_in_China

115.   Everhard J. Slijper, *Riesen des Meeres; eine Biologie der Wale und Delphine* (Berlin: Springer-Verlag, 1962) 15.

        Translated by John Drury, *Whales and Dolphins* (Ann Arbor: University of Michigan Press, 1976) 18.

116.   Raymond Sutera, "The Origin of Whales and the Power of Independent Evidence," The TalkOrigins Archive, 2001,
        http://www.talkorigins.org/features/whales

See also:

Wikipedia, "Evolution of cetaceans,"

http://en.wikipedia.org/wiki/Evolution_of_cetaceans

117. BBC News, "Bee fossil, DNA generate a buzz," 25 October 2006,
      http://news.bbc.co.uk/2/hi/science/nature/6084974.stm

      See also:

      George O. Poinar et al., "A Fossil Bee from Early Cretaceous Burmese
      Amber," *Science* Vol. 314, No. 5799, 27 October 2006: 614,
      http://www.sciencemag.org/content/314/5799/614.abstract

118. Wikipedia, "Flowering plant," Evolution,
      http://en.wikipedia.org/wiki/Flowering_plant#Evolution

119. Edward B. Daeschler et al., "A Devonian tetrapod-like fish and the
      evolution of the tetrapod body plan," *Nature* 440, 6 April 2006: 757-63,
      http://www.nature.com/nature/journal/v440/n7085/full/nature04639
      .html

      See also:

      University of Chicago, "*Tiktaalik roseae*,"
      http://tiktaalik.uchicago.edu

120. Darwin, *The Origin of Species*, 484.

121. Wikipedia, "Ole Roemer,"
      http://en.wikipedia.org/wiki/Ole_Roemer

122. Wikipedia, "Antarctic Circumpolar Current,"
      http://en.wikipedia.org/wiki/Antarctic_Circumpolar_Current

123. William Shakespeare, *Romeo And Juliet*, II, ii, 2-3,
      http://shakespeare.mit.edu/romeo_juliet/full.html

124. Philip E. Johnson, "Starting a Conversation about Evolution. A review
      of *The Battle of the Beginnings: Why Neither Side is Winning the
      Creation-Evolution Debate* by Del Ratzsch," August 31, 1996,
      http://www.arn.org/docs/johnson/ratzsch.htm

      See also:

      Forrest, *Creationism's Trojan Horse*, 314.

125. The quote is attributed to Swiss theologian Frédéric Louis Godet
      (continued)

(1812-1900).

See for example:
Bible Study Tools, "Romans – Introduction," 2010,
http://www.biblestudytools.com/commentaries/peoples-new-testa
ment/romans/introduction.html

126.   Cornelia Dean, "Scientists Speak Up on Mix of God and Science," *The New York Times*, August 23, 2005,
http://www.nytimes.com/2005/08/23/national/23believers.html

Stefan Lovgren, "Evolution and Religion Can Coexist, Scientists Say,"
National Geographic News, October 18, 2004,
http://news.nationalgeographic.com/news/2004/10/1018_041018_sci
ence_religion.html

127.   University of California Museum of Paleontology – Understanding
Evolution, "Lines of evidence: The science of evolution,"
http://evolution.berkeley.edu/evolibrary/article/_0_0/lines_01

128.   Acts 5:10

129.   Philip Johnson, American Family Radio, Jan. 10, 2003,
http://en.wikiquote.org/wiki/Phillip_E._Johnson

130.   Astronomy Picture of the Day, "M7: Open Star Cluster in Scorpius,"
November 8, 2009,
http://antwrp.gsfc.nasa.gov/apod/ap091108.html

131.   A few of the names of God from *Genesis 17:1, Psalms 136:2, Isaiah 40:28, Daniel 7:9,* and *Revelation 1:17*

Bibliography

"Analysis of the Discovery Institute's Bibliography." National Center for Science, June 1, 2002.
http://ncseweb.org/creationism/general/analysis-discovery-institutes-bibliog raphy

"Antarctic Circumpolar Current." Wikipedia.
http://en.wikipedia.org/wiki/Antarctic_Circumpolar_Current

"*Archaeopteryx*: An Early Bird." University of California Museum of Paleontology.
http://www.ucmp.berkeley.edu/diapsids/birds/archaeopteryx.html

"Ashkenazi Jews in Early Modern Europe." European History Online.
http://www.ieg-ego.eu/en/threads/europe-on-the-road/jewish-migration/ash kenazi-jews-in-early-modern-europe/predrag-bukovec-ashkenazi-jews-in-ea rly-modern-europe

"*Australopithecus afarensis*." Wikipedia.
http://en.wikipedia.org/wiki/Australopithecus_afarensis

"Bacterial Generation time and example of fast growing bacteria and slow growing bacteria?" WikiAnswers.
http://wiki.answers.com/Q/Bacterial_Generation_time_and_example_of_fas t_growing_bacteria_and_slow_growing_bacteria

"Basel." The Black Death. The Jewish Encyclopedia.
http://www.jewishencyclopedia.com/view.jsp?artid=375&letter=B&search=B asel

"Basel massacre." Wikipedia.
http://en.wikipedia.org/wiki/Basel_massacre

"Basel: some history." Switzerland Is Yours.
http://switzerland.isyours.com/e/guide/basel/history.html

"Bee fossil, DNA generate a buzz." *BBC News*, 25 October 2006.
http://news.bbc.co.uk/2/hi/science/nature/6084974.stm

Behe, Michael J. *Darwin's Black Box: The Biochemical Challenge to Evolution*. New York: Touchstone, 1996.

Ben-Sasson, H. H., ed. *A History of the Jewish People*. Cambridge: Harvard University Press, 1976.

"Black Death." The Jewish Encyclopedia.
http://www.jewishencyclopedia.com/view.jsp?artid=1114&letter=B&search= BlackDeath

"BLACK PLAGUE (Europe), 1348 - 1349." The History of the Jewish People.
http://www.jewishhistory.org.il/history.php?startyear=1340&endyear=1349
Britt, Robert Roy. "Scientists' Belief in God Varies Starkly by Discipline."
LiveScience, August 11, 2005.
http://www.livescience.com/strangenews/050811_scientists_god.html
Carrier, Richard. "Addendum B: Are the Odds Against the Origin of Life Too Great
to Accept?" The Secular Web.
http://www.infidels.org/library/modern/richard_carrier/addendaB.html
"Christian Clergy Letter." The Clergy Letter Project.
http://theclergyletterproject.org/
Comparative Study Bible. Grand Rapids: Zondervan, 1984.
Connor, Steve. "Moth study backs classic 'test case' for Darwin's theory." The
Independent, August 25, 2007.
http://freethoughtpedia.com/wiki/Peppered_Moth
"Crusades, The." First Crusade: 1096. The Jewish Encyclopedia.
http://www.jewishencyclopedia.com/view.jsp?artid=908&letter=C&search=
Crusades
Daeschler, Edward B. et al. "A Devonian tetrapod-like fish and the evolution of the
tetrapod body plan." Nature 440, 6 April 2006: 757-63.
http://www.nature.com/nature/journal/v440/n7085/full/nature04639.html
Darwin, Charles R. On the Origin of Species by Means of Natural Selection, or the
Preservation of Favoured Races in the Struggle for Life. London: John Murray, 1859.
Darwin, Charles R. The Complete Works of Charles Darwin Online.
http://darwin-online.org.uk
Darwin, Charles R. The Descent of Man, and Selection in Relation to Sex. London: John
Murray, 1871.
Darwin, Charles R. Voyages of the Adventure and Beagle, Vol. III. London: Henry
Colburn, 1839.
Dean, Cornelia. "Scientists Speak Up on Mix of God and Science." The New York
Times, August 23, 2005.
http://www.nytimes.com/2005/08/23/national/23believers.html
Dembski, William A. The Design Inference: Eliminating Chance Through Small
Probabilities. Cambridge: Cambridge University Press, 1998.
Dembski, William A. Intelligent Design: The Bridge Between Science & Theology.
Downers Grove: InterVarsity Press, 1999.
"Diagram from the Origin of Species." The Friends of Charles Darwin.
http://friendsofdarwin.com/docs/origin-1/diagram
Dickerson, David M. "Clifford's Tower: Massacre at York (1190)." 2 July 1997.
http://ddickerson.igc.org/cliffords-tower.html
Doolittle, Russell F. "A Delicate Balance." Boston Review, February/March 1997.
http://bostonreview.net/BR22.1/doolittle.html
Drake, Finis Jennings. Drake's Annotated Reference Bible. Lawrenceville: Drake Bible
Sales, 1991.

BIBLIOGRAPHY

Dubnow, Simon M. *History of the Jews in Russia and Poland*. Trans. I. Friedlander. Philadelphia: The Jewish Publication Society of America, 1918.

Ekstrand, Donald W. *Christianity: The Pursuit of Divine Truth*. Xulon Press, 2008.

"Emperor of Japan." Wikipedia.
http://en.wikipedia.org/wiki/Emperor_of_Japan

"England." The Expulsion. The Jewish Encyclopedia.
http://www.jewishencyclopedia.com/view.jsp?artid=375&letter=E&search=England#957

"*Ensatina eschscholtzi*, Speciation in Progress." *Ensatina* web site.
http://www.ensatina.net

"Evidence For Evolution." All About Creation.
http://www.allaboutcreation.org/evidence-for-evolution.htm

"Evolution of cetaceans." Wikipedia.
http://en.wikipedia.org/wiki/Evolution_of_cetaceans

Expelled: No Intelligence Allowed [Motion picture]. Ruloff, W. (Producer), & Frankowski, N. (Director). United States: Premise Media Corporation, 2008.

"Fahrenheit 9/11." Wikipedia.
http://en.wikipedia.org/wiki/Fahrenheit_9/11

Fidelibus, Alex. "The Evolution of the Mousetrap through Natural Selection."
http://www.fidelibus.com/mousetrap

"Flowering Plant," Evolution. Wikipedia.
http://en.wikipedia.org/wiki/Flowering_plant#Evolution

Forrest, Barbara and Paul R. Gross. *Creationism's Trojan Horse, The Wedge of Intelligent Design*. New York: Oxford University Press, 2005.

Foster, David. *The Philosophical Scientists*. New York: Barnes & Noble, 1993.

"Fourth Lateran Council (1215)." New Advent.
http://www.newadvent.org/cathen/09018a.htm

Frack, Donald. "Peppered Moths - round 2 (part 2 of 2)." April 5, 1999.
http://www.asa3.org/archive/evolution/199904/0103.html

"France." Exile of 1306. The Jewish Encyclopedia.
http://www.jewishencyclopedia.com/view.jsp?artid=288&letter=F&search=France#820

Frank, Mark. "Detecting design: Specification versus Likelihood." Talk Reason, July 2, 2006.
http://www.talkreason.org/articles/likely.cfm

Gaebelein, Frank E., and J. D. Douglas, eds. *The Expositor's Bible Commentary*. Grand Rapids: Zondervan, 1979.

Gishlick, Alan D. "Icons of Evolution? Why much of what Jonathan Wells writes about evolution is wrong." National Center for Science Education, October 19, 2008.
http://ncseweb.org/creationism/analysis/icons-evolution

"Godfrey of Bouillon." Wikipedia.
http://en.wikipedia.org/wiki/Godfrey_of_Bouillon

Gould, Stephen Jay. "Impeaching a Self-Appointed Judge." *Scientific American,* July 1992: 118-21.
    http://www.stephenjaygould.org/ctrl/gould_darwin-on-trial.html

"Guillermo Gonzalez." Expelled Exposed: The Truth behind the Fiction. National Center for Science Education.
    http://www.expelledexposed.com/index.php/the-truth/gonzalez

Halsall, Paul. "Medieval Sourcebook: Ephraim of Bonn: The York Massacre 1189-90." Internet Medieval Sourcebook, Fordham University.
    http://www.fordham.edu/halsall/source/ephr-bonn1.asp

Halsall, Paul. "The Black Death and the Jews 1348-1349 CE." Jewish History Sourcebook, Fordham University.
    http://www.fordham.edu/halsall/jewish/1348-jewsblackdeath.html

Hardy, David. "Michael Moore Exposed,"
    http://www.mooreexposed.com

Hitler, Adolf. *Mein Kampf.* Boston: Houghton Mifflin, 1999.

Hitler, Adolf. *Mein Kampf.* New York: Reynal And Hitchcock, 1941.
    http://archive.org/details/meinkampf035176mbp

Holy Bible, The New International Version. Grand Rapids: Zondervan, 1978.

Holy Bible, The New King James Version. Thomas Nelson, Inc., 1984.

Ho-Stuart, Chris. "Irreducible Complexity as an Evolutionary Prediction." The TalkOrigins Archive, Post of the Month, September 2006.
    http://www.talkorigins.org/origins/postmonth/sep06.html

Hoyle, Fred. *The Intelligent Universe: A New View of Creation and Evolution.* New York: Holt, Rinehart and Winston, 1984.

Hoyle, Fred and N. C. Wickramasinghe. *Evolution From Space: A Theory of Cosmic Creation.* New York: Simon & Schuster, 1981.

Hyun, Jung-Ho and Eun-Jin Yang. "Freezing Seawater for the Long-term Storage of Bacterial Cells for Microscopic Enumeration." *The Journal of Microbiology,* Vol. 41, No. 3, September 2003: 262-5.
    http://www.msk.or.kr/jsp/downloadPDF1.jsp?fileName=413-15.pdf

"Jim Jones." Wikipedia.
    http://en.wikipedia.org/wiki/Jim_Jones

Johnson, Philip E. American Family Radio, January 10, 2003.
    http://en.wikiquote.org/wiki/Phillip_E._Johnson

Johnson, Philip E. *Darwin on Trial.* Downers Grove: InterVarsity Press, 2nd edition, 1993.

Johnson, Philip E. "Starting a Conversation about Evolution. A review of *The Battle of the Beginnings: Why Neither Side is Winning the Creation-Evolution Debate* by Del Ratzsch." August 31, 1996.
    http://www.arn.org/docs/johnson/ratzsch.htm

"Jonathan Wells (intelligent design advocate)." Wikipedia.
    http://en.wikipedia.org/wiki/Jonathan_Wells_(intelligent_design_advocate)

Kelland, Kate. "HIV/AIDS drug puzzle cracked." *Reuters,* February 1,2010.
    http://www.reuters.com/article/healthNews/idUSTRE6101AQ20100201

Keller, Greta et al. "New evidence concerning the age and biotic effects of the Chicxulub impact in NE Mexico." *Journal of the Geological Society* Vol. 166, Part 3, May 2009: 393-411.
  http://jgs.geoscienceworld.org/cgi/content/abstract/166/3/393
Kitzmiller v. Dover Area School Dist. 400 F.Supp.2d 707, 741 (M.D.Pa. 2005).
Kitzmiller v. Dover Area School District. "Trial transcript: Day 12 (October 19, 2005), AM Session, Part 2." The TalkOrigins Archive.
  http://www.talkorigins.org/faqs/dover/day12am2.html
Kitzmiller v. Dover Area School District. "Trial transcript: Day 20 (November 3, 2005), PM Session, Part 2." The TalkOrigins Archive.
  http://www.talkorigins.org/faqs/dover/day20pm2.html
Kitzmiller v. Dover Area School District. "Trial transcript: Day 21 (November 4, 2005), AM Session, Part 2." The TalkOrigins Archive.
  http://www.talkorigins.org/faqs/dover/day21am2.html
Kreis, Steven. "Lecture 29 – Satan Triumphant: The Black Death." The History Guide: Lectures on Ancient and Medieval European History, August 03, 2009.
  http://www.historyguide.org/ancient/lecture29b.html
"Laws of thermodynamics." Wikipedia.
  http://en.wikipedia.org/wiki/Laws_of_thermodynamics
"Life." Wikipedia.
  http://en.wikipedia.org/wiki/Life#Definitions
"Lines of evidence: The science of evolution." Understanding Evolution.
  http://evolution.berkeley.edu/evolibrary/article/_0_0/lines_01
Lovgren, Stefan. "Evolution and Religion Can Coexist, Scientists Say." National Geographic News, October 18, 2004.
  http://news.nationalgeographic.com/news/2004/10/1018_041018_science_religion.html
Luther, Martin. *On the Jews and Their Lies.* Fortress Press & Augsburg Fortress, 1971.
  http://www.humanitas-international.org/showcase/chronography/documents/luther-jews.htm
Maddison, D. R. and K. S. Schulz (Editors). The Tree of Life Web Project. 2007.
  http://tolweb.org/tree
Majerus, Michael E. N. *Melanism: Evolution in Action.* Oxford: Oxford University Press, 1998.
Majerus, Michael E. N. *Moths.* London: HarperCollins Publishers, 2002.
"'Masterful' Federal Ruling on Intelligent Design Was Copied from ACLU." Discovery Institute, December 12, 2006.
  http://www.discovery.org/a/3828
Mathis, Mark. "A Conversation with *Expelled*'s Associate Producer Mark Mathis." *Scientific American*, April 9, 2008: at 18 minutes, 54 seconds.
  http://www.scientificamerican.com/article.cfm?id=a-conversation-with-mark-mathis
Matzke, Nick. "Icon of Obfuscation." The TalkOrigins Archive, 2002.
  http://www.talkorigins.org/faqs/wells/iconob.html

McDonald, John H. "A Reducibly Complex Mousetrap." 2002.
   http://udel.edu/~mcdonald/mousetrap.html
"Mexican tetra." Wikipedia.
   http://en.wikipedia.org/wiki/Mexican_tetra
Meyer, Stephen C. "Teach the Controversy on Origins." *Cincinnati Enquirer*,
   Commentary Section, March 30, 2002.
   http://www.arn.org/docs/meyer/sm_teachthecontroversy.htm
Miller, Kenneth R. "A Review of *Darwin's Black Box*." *Creation Evolution Journal* 16,
   1996: 36-40.
   http://www.millerandlevine.com/km/evol/behe-review/index.html
Mitchell, Tommy. "Much Ado About Moths." Answers in Genesis, April 23, 2008.
   http://www.answersingenesis.org/articles/aid/v3/n1/much-ado-about-moths
Moreland, James Porter (Editor). *The Creation Hypothesis: Scientific Evidence for an
   Intelligent Designer*. InterVarsity Press, Downers Grove, IL, 1994.
Morris, Henry M. *The Defender's Study Bible, King James Version*. Grand Rapids:
   World Publishing, Inc., 1995.
Morris, Henry M. *The Remarkable Birth of Planet Earth*. Bethany Fellowship, Inc.,
   Minneapolis, 1972.
Muller, Hermann J. "Genetic variability, twin hybrids and constant hybrids, in a
   case of balanced lethal factors." *Genetics* 3, 1918: 464.
   http://www.genetics.org/cgi/reprint/3/5/422
"M7: Open Star Cluster in Scorpius." Astronomy Picture of the Day. November 8,
   2009.
   http://antwrp.gsfc.nasa.gov/apod/ap091108.html
"Natural nuclear fission reactor." Wikipedia.
   http://en.wikipedia.org/wiki/Natural_nuclear_fission_reactor
Nave's Compact Topical Bible. Grand Rapids: Zondervan, 1972.
Nedin, Chris. "All About *Archaeopteryx*." The TalkOrigins Archive, 1999.
   http://www.talkorigins.org/faqs/archaeopteryx/info.html
New American Standard Bible, Side Margin Reference Edition with Concordance.
   Thomas Nelson Publishers, New York, 1977.
"New Species." The TalkOrigins Archive, Index to Creationist Claims. Claim CB910,
   ed. Mark Isaak, 2007.
   http://www.talkorigins.org/indexcc/CB/CB910.html
Noebel, David A. *Understanding the Times: The Religious Worldviews of Our Day and
   the Search for Truth*. Harvest House Publishers, 1st Edition, 1991.
"Objections to the Doctrine of Evolution." The Bible Probe.
   http://bibleprobe.org/objection.html
Okoh, Anthony I., et al. "Aerobic Heterotrophic Bacterial and Fungal Communities
   in the Topsoil of Omo Biosphere Reserve in Southwestern Nigeria." *Biotropica*
   Vol. 32, Issue 2, June 2000: 208-12.
   http://onlinelibrary.wiley.com/doi/10.1111/j.1744-7429.2000.tb00463.x/abstract
"Ole Roemer." Wikipedia.
   http://en.wikipedia.org/wiki/Ole_Roemer

"Origin of birds," Modern research and feathered dinosaurs in China. Wikipedia.
http://en.wikipedia.org/wiki/Dinosaur-bird_connection#Modern_research_a
nd_feathered_dinosaurs_in_China

Patterson, Colin. *Evolution*. Ithaca: Cornell University Press, 2nd edition, 1999.

"Peppered moth evolution." Wikipedia.
http://en.wikipedia.org/wiki/Peppered_moth_evolution

Perakh, Mark. "A Militant Dilettante in Judgement of Science, How a lawyer (Phillip Johnson) disproves Darwinism." Talk Reason, 2000.
http://www.talkreason.org/articles/johnson.cfm

Perloff, James. *Tornado in a Junkyard: The Relentless Myth of Darwinism*. Refuge Books, Arlington, MA, 1999.

Poinar, George O. et al. "A Fossil Bee from Early Cretaceous Burmese Amber." *Science* Vol. 314, No. 5799, 27 October 2006: 614.
http://www.sciencemag.org/content/314/5799/614.abstract

"Prediction," Prediction in Science. Wikipedia.
http://en.wikipedia.org/wiki/Prediction_in_science#Prediction_in_science

"Pulsar." Wikipedia.
http://en.wikipedia.org/wiki/Pulsar#Discovery

Rana, Fazale R. "Tetrapod Transitions: Evidence for Design." *New Reasons To Believe* Vol. 1, No. 1, 2009: 6-7.
http://www.reasons.org/articles/tetrapod-transitions-evidence-for-design

"Richard Sternberg." Expelled Exposed: The Truth behind the Fiction. National Center for Science Education.
http://www.expelledexposed.com/index.php/the-truth/sternberg

"Romans – Introduction." Bible Study Tools, 2010.
http://www.biblestudytools.com/commentaries/peoples-new-testament/rom
ans/introduction.html

Roth, Cecil, and Geoffrey Wigoder, eds. *Encyclopaedia Judaica*. Jerusalem: Keter Publishing House, Ltd., 1972.

Roussel, Erwan G., et al. "Extending the Sub-Sea-Floor Biosphere." *Science* Vol. 320. No. 5879, 23 May 2008: 1046.
http://www.sciencemag.org/content/320/5879/1046.abstract

Schrier, Stanley L. MD and Stephen A. Landaw MD, PhD. "Introduction to hemoglobin mutations." UpToDate.
http://www.uptodate.com/patients/content/topic.do?topicKey=~OrEJEbFF8
hpe85O

Schulte, Peter et al. "The Chicxulub Asteroid Impact and Mass Extinction at the Cretaceous-Paleogene Boundary." *Science*, Vol. 327. no. 5970, March 5, 2010: 1214-8.
http://www.sciencemag.org/content/327/5970/1214.abstract

"Scientific method," Characterizations, Definition. Wikipedia.
http://en.wikipedia.org/wiki/Scientific_method#Characterizations

Scott, Eugenie C. "A Rude Introduction to "Expelled"." Reports of the National

(*continued*)

Center for Science Education, Vol. 28. Issue 5-6, September-December, 2008: 11-2.
  http://ncseweb.org/rncse/28/5-6/rude-introduction-to-expelled
Scott, Eugenie C. "Darwin Prosecuted: Review of Johnson's Darwin on Trial." Creation Evolution Journal, Vol. 13., No. 2, Winter 1993.
  http://ncseweb.org/cej/13/2/darwin-prosecuted-review-johnsons-darwin-trial
Search The Bible. BibleStudyTools.
  http://www.biblestudytools.com
Shakespeare, William. *Romeo And Juliet.*
  http://shakespeare.mit.edu/romeo_juliet/full.html
Slijper, Everhard J. *Riesen des Meeres; eine Biologie der Wale und Delphine.* Berlin: Springer-Verlag, 1962.
Slijper, Everhard J. *Whales and Dolphins.* Translated by John Drury. Ann Arbor: University of Michigan Press, 1976.
Stott, Philip. *Vital Questions.* Reformation Media & Press, 2nd edition, 2002.
Strong, James. *The Exhaustive Concordance of The Bible.* Abingdon Press, 1961.
Sutera, Raymond. "The Origin of Whales and the Power of Independent Evidence." The TalkOrigins Archive, 2001.
  http://www.talkorigins.org/features/whales
Sweeney, John and Bill Law. "Gene find casts doubt on double 'cot death' murders." The Observer, July 15, 2001.
  http://www.guardian.co.uk/uk/2001/jul/15/johnsweeney.theobserver
Tackett, Del. *The Truth Project DVD Set.* Colorado Springs: Focus On The Family, 2006. Lesson 5b, Slide #16.
"That quote! — about the missing transitional fossils." Creation Ministries International.
  http://creation.com/that-quoteabout-the-missing-transitional-fossils
Theobald, Douglas. "The Mullerian Two-Step: Add a part, make it necessary or, Why Behe's 'Irreducible Complexity' is silly." The TalkOrigins Archive, July 18, 2007.
  http://www.talkorigins.org/faqs/comdesc/ICsilly.html
Theunissen, Lionel. "Patterson Misquoted: A Tale of Two 'Cites'." The TalkOrigins Archive, 1997.
  http://www.talkorigins.org/faqs/patterson.html
"*Tiktaalik roseae.*" University of Chicago.
  http://tiktaalik.uchicago.edu
"Traditional peppered moth story is no longer supportable, The." The TalkOrigins Archive, Index to Creationist Claims. Claim CB601, ed. Mark Isaak, 2005.
  http://www.talkorigins.org/indexcc/CB/CB601.html
Transitional Fossils? Evolutionists Say No! YouTube.
  http://www.youtube.com/watch?v=typrxYyiEEI
Unlocking the Mysteries of Life [Video]. Adams, James W. (Producer) & Allen, Lad (Director). United States: Illustra Media, 2003.
  http://www.youtube.com/watch?v=VWvS1UfXl8k

174

BIBLIOGRAPHY

"Ursid hybrid." Wikipedia.
 http://en.wikipedia.org/wiki/Ursid_hybrid
U.S. District Court for the Middle District of Pennsylvania. Memorandum Opinion
 December 20, 2005: 78, 137-8.
 http://www.pamd.uscourts.gov/kitzmiller/kitzmiller_342.pdf
Wells, Jonathan. "Darwinism: Why I Went for a Second Ph.D." True Parents
 Organization, The Words of the Wells Family.
 http://www.tparents.org/library/unification/talks/wells/DARWIN.htm
Wells, Jonathan. *Icons of Evolution: Science or Myth? Why Much of What We Teach
 About Evolution Is Wrong.* Washington, DC: Regnery Publishing, 2000.
West, John G. and David K. DeWolf. "A Comparison of Judge Jones' Opinion in
 Kitzmiller v. Dover with Plaintiffs' Proposed 'Findings of Fact and Conclusions
 of Law'." Discovery Institute, December 12, 2006.
 http://www.discovery.org/scripts/viewDB/filesDB-download.php?comman
 d=download&id=1186
"What has the head of a crocodile and the gills of a fish?" Understanding Evolution,
 June 2009.
 http://evolution.berkeley.edu/evolibrary/news/060501_tiktaalik
"Wright Brothers." Wikipedia.
 http://en.wikipedia.org/wiki/Wright_brothers
Wright, Orville. *How We Invented The Airplane.* New York: Dover Publications, 1988.
Young, Matt. "Dembski's Explanatory Filter Delivers a False Positive." The Panda's
 Thumb, April 22, 2004.
 http://pandasthumb.org/archives/2004/04/dembskis-explan.html
"ZHITOMIR (Russia), 1753 May 26." This Day In Jewish History.
 http://www.jewishhistory.org.il/history.php?startyear=1750&endyear=1759

Index

www.ingramcontent.com/pod-product-compliance
Lightning Source LLC
Chambersburg PA
CBHW030416100426
42812CB00028B/2987/J